Legal Disclaimer

Everything Electrical: How to Use All The Functions On Your Multimeter (Book One)

Preface:

Have you ever studied electricity in high school, a college class or maybe a trade school and felt like the teacher didn't tell you everything you needed to know? Or that the theory just didn't give you anything useful to use out on the field? That they didn't prepare you for those uncommon or intermittent electrical issues that leave you feeling like you don't have a plan of attack. Well either way GREAT! You are not alone. I myself read at least 10 full textbooks on electrical, electronics, industrial electrical and automotive electricity, that by the way were not very cheap averaging in cost around 150$ each. But these books still left me feeling like they failed in many aspects for learning real world electrical tips and tricks.

This book was written to educate in a simpler way for everyone to understand, beginners and veteran technicians alike. There is no reason to complicate things with big words that usually are left unexplained by other books and make it even harder to understand with bad examples. This book is priced low but because I feel that everyone should know at least the basics. I will include many examples of each topic I discuss for better understanding. Because of my approach to certain topics, I recommend that you read the book front to back even if you feel you've already read too much theory of electricity. My goal is to make you "the electrical guy" that will fearlessly tackle any job. If this book series "Everything Electrical" does not teach you everything you wanted to know, I guarantee that it will at least be a very powerful supplement to your learning of electrical testing at a low price. Thank you and I hope you enjoy.

My Own Take on Electricity

Everyone should know that anything technical, including electricity, involves a lot of complicated physics and if I wanted to I can talk all day about how things really happen down to the atomic level. But for the sake of keeping it simple and teaching only what you need to know to get to working on electrical problems as soon as possible, just accept that the way that I'm explaining the topics in this book is for your ease. My methods and examples do teach and WILL work in the real world for real life electrical issues.

This book is the first book of the "Everything Electrical" series that will aim to teach you how you too can use your electrical test meters like a pro.

Table Of Contents:

Ch. 1: Introduction To Multimeters: Manual, Auto Ranging and Analog

A multi-meter is an electrical testing tool that combines many different kinds of electrical meters into one tool. The position of the dial on the multimeter will indicate what meter setting you are currently on. Each of the symbols on the multimeter represents a different **unit of measure** that measures something entirely different from the other dial settings.

SYMBOLS	DEFINITIONS
DCV, DC, $\overline{V}$, $m\overline{V}$, $u\overline{V}$	-DIRECT VOLTAGES (DC VOLTS)
ACV, AC, $\tilde{V}$, $m\tilde{V}$, $u\tilde{V}$	-ALTERNATING VOLTAGES (AC VOLTS)
Ω	-RESISTANCE, OHMS
•)))	-CONTINUITY TESTER
—▶|—	-DIODE TESTER
—)|—	-CAPACITANCE TESTER
$\overline{A}$, DCA, $m\overline{A}$, $u\overline{A}$	-DIRECT AMPERAGE (DC) CURRENT FLOW (DC) AMPS
$\tilde{A}$, ACA, $m\tilde{A}$, $u\tilde{A}$	-ALTERNATING AMPERAGE (AC) CURRENT FLOW (AC) AMPS
Hz, FREQ	-FREQUENCY TESTER (HERTZ)
DUTY CYCLE	-DUTY CYCLE TEST
TEMP, C, F, K	-TEMPERATURE MEASURE

(This chart shows the definitions to all the common units of measure (or meter settings) often seen around a multimeter's dial. Each dial setting is measuring something different and is represented by its own unique symbol. The three most commonly known units of measure are Voltage, Amperage and Resistance)

There are also three main types of multi-meters that are currently in the market today. The Analog multi-meter, the Manual Ranging multi-meter and the Auto-Ranging multi-meter. The way to perform a test using either of these multi-meters is basically the same. The only difference between the meters is in the setup before testing.

Let's take a look at the faces of each of the multi-meters to see what they commonly look like and also note all the important factors about their setup.

Analog Multimeters:

The Analog multi-meter was the first type of multimeter to come out many decades ago. As with many other early equipment, this meter required more skill to setup and better understanding of the symbols on the meter.

The symbols around the multimeter dial represent the units of measurement that are available on the meter.

(This image shows a common analog multimeter. As shown, there are four main units of measure on this specific analog meter. The units are DC volts, AC volts, Resistance/Ohms and DC Amperage)

Each unit of measurement on the meter dial then has sub-settings under it that are represented by either numbers or other symbols. When the sub-setting is a number, it is simply one of the many ranges available for that specific unit. If the sub-setting is a symbol, this indicates other testing options available within that unit. To explain this further we will explore the unit ranges that are available on an analog multimeter.

Unit Ranges:

A unit range is best defined as the maximum number value that you can measure when having the meter dial set to that specific unit range setting. For example, in the 10 volt range setting, the meter will only be able to measure up to 10 volts. If during testing you are measuring anything that is over 10 volts, the meter will either display an over-limit (OL) reading or simply nothing at all. This type of multimeter requires that you know more or less how much of the unit you will expect to measure in order to have the proper setup before you begin measurement.

To set the meter up for testing, move the dial to the unit you want to measure in as well as the expected range you plan to use.

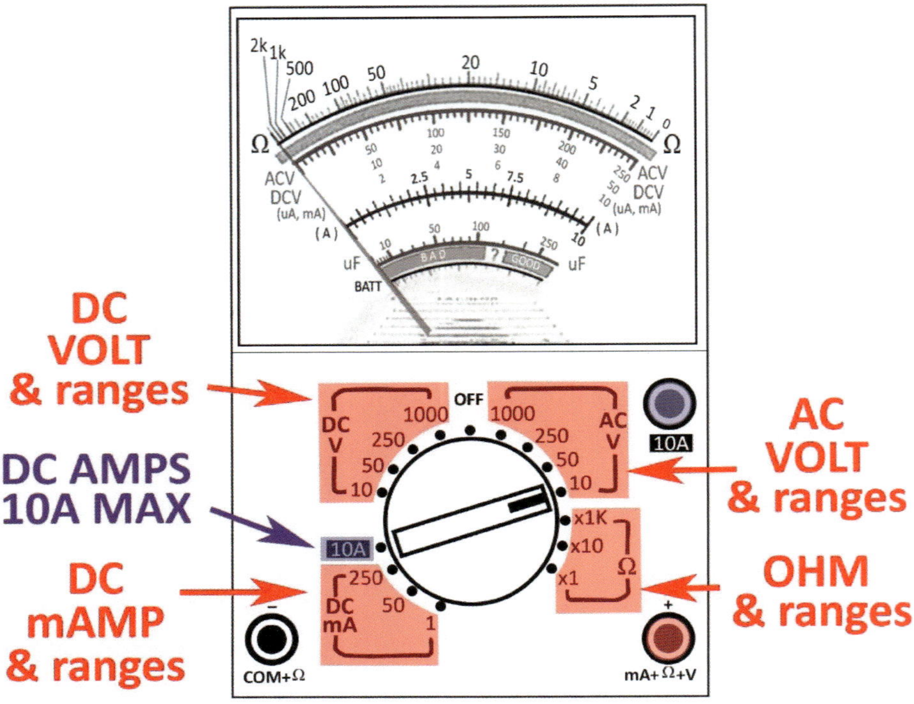

(This image shows the basic analog multimeter with the units and unit ranges highlighted for easy identification. The dial position on the meter is set to the unit for AC Volts and the unit range of 10 volts. This setup will allow you to measure any AC voltage up to 10 volts.)

As you can see around the dial of the meter there are many units and unit ranges to choose from. The capital "DCV" symbol is a unit called Direct Volts or DC volts and has many different number ranges under it. It includes the 10 volt, 50 volt, 250 volt and 1000 volt ranges. All these ranges belong under one unit of measure, the DC volt.

Another example is the unit "mA" or milliamps. This unit has dial ranges that include the 1mA, 50mA and 250mA ranges. All these ranges belong under one unit of measure, the milliamp.

A third example is in the unit for ohms otherwise known as resistance. If you notice, the number ranges under "ohms" has an X before the number. The dial ranges are x1, x10 and x1k. This simply means that whatever value you read on the meter display will then have to be multiplied by the unit range selected. For example, if you read 5 ohms on the meter screen and you are on the x1k unit range the true measurement value is 5000 ohms. (5 **x 1k**= 5000).

Tip for Selecting Ranges: If you own one of these meters or a similar meter that requires you to manually select ranges, the best way to use these without having to worry so much about what unit ranges to choose is to start at the highest range first. If when measuring, the number value measured on the meter display is too small to read properly you should move down one range for a better resolution.

The Meter Display:

The meter display on an analog multimeter is represented by the many unit symbols (that are also on the dial of the meter) each with its own number line. The position of the needle on the meter screen will indicate the measurement value during testing. One thing to understand about the various number lines on the screen is that you must only pay attention to the number line that belongs to the symbol that you have chosen on the meter dial.

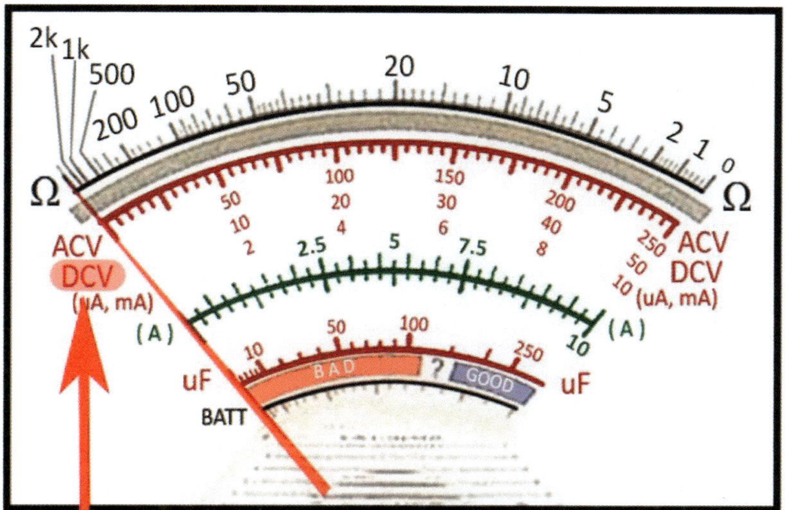

If the Dial setting chosen is under DC Volts, You must only pay attention to the numbers that are next to the DCV scale on the meter display.

(This image represents how to read the measurement display on an analog multimeter.)

After you have chosen your unit and range using the meter dial, the next step is to pay attention to only the number values that match the unit selected. For example, the red number line on the upper part of the screen belongs to the ACV, DCV and the uA,mA symbol. When either of these units of measures are selected on the dial you should only pay attention to the values belonging to the red number line next to the symbol.

Proper Test Probe Placement:

The final step when setting up your meter is to connect the test probe into their proper meter socket. Included with the meter will come a set of regular test probes. One black test probe, one red colored test probe.

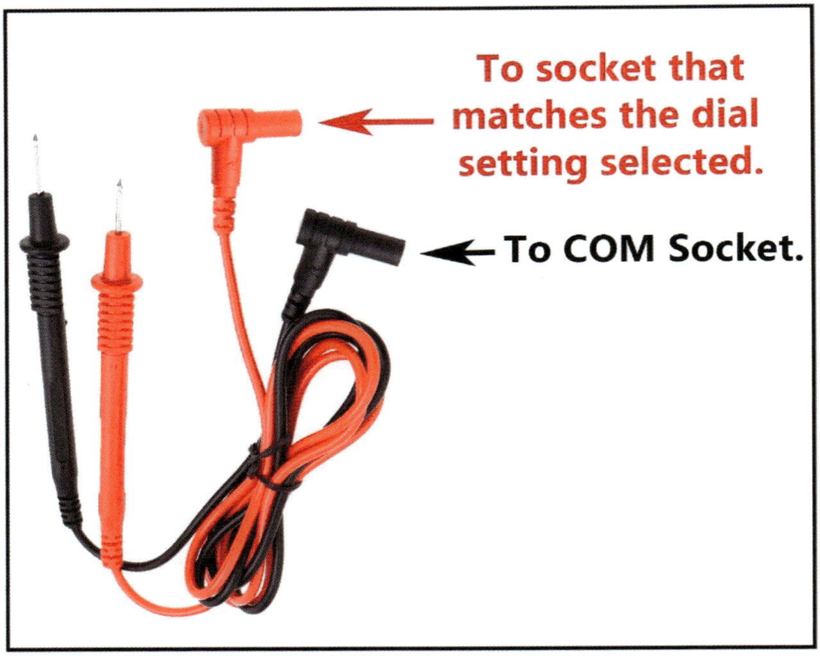

(This image shows the test probes that will also come along with your purchase of a multimeter.)

The black test probe will always be installed in the COM or common socket of the meter, regardless of what dial setting you are on. The red probe on the other hand, will have to be moved around to a different socket depending on what dial setting you selected. The red probe is installed in the socket that matches the symbol you have selected with the dial on your meter. For example, if the dial is on the Amperage setting or "A" symbol, you will install the red probe into the socket that has the same Amperage symbol as the one on the dial setting you selected.

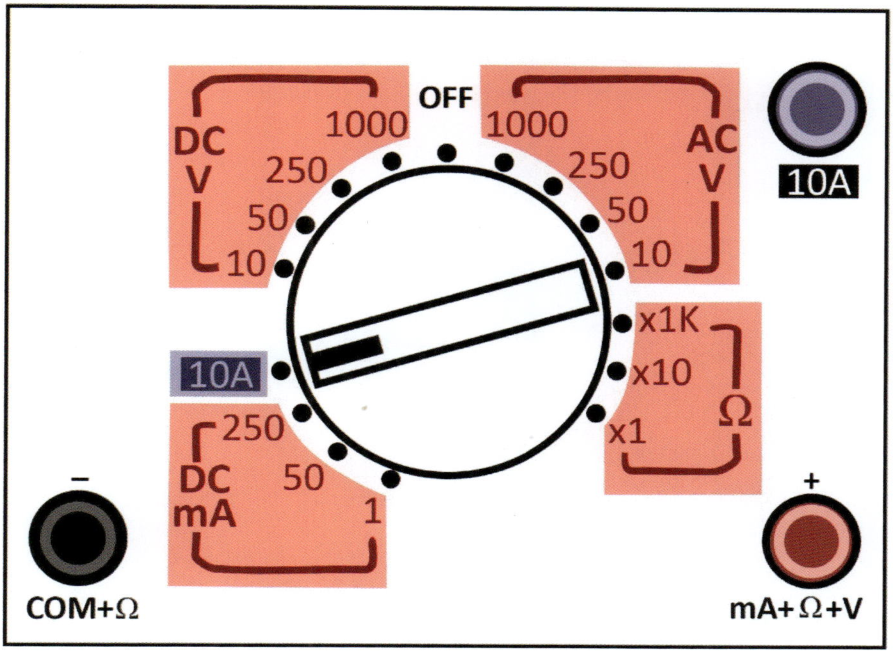

(In this image we show the meter dial settings highlighted to match their respective meter socket. The black test probe will always be installed in the socket labeled COM. The red test probe will be installed in the socket whose label matches the symbol on the dial setting that was selected.)

The dial setting selected was the "10A" symbol so the red probe will be installed in the socket with the same 10A symbol. After the probes are installed, the multimeter is now set up and ready to begin test measurements.

Using the Analog Meter:

After you have selected your unit of measure, unit range and have installed the test probes in the proper socket, the meter is now ready to perform a measure for the unit selected.

(In this illustration we see an example of the analog meter fully setup to test DC Voltage and connected to terminals of a 12v battery. The unit range for this setup is 50 max DC Volts. The meter is connected to read the voltage available inside this power source. The meter needle reads 12 volts.)

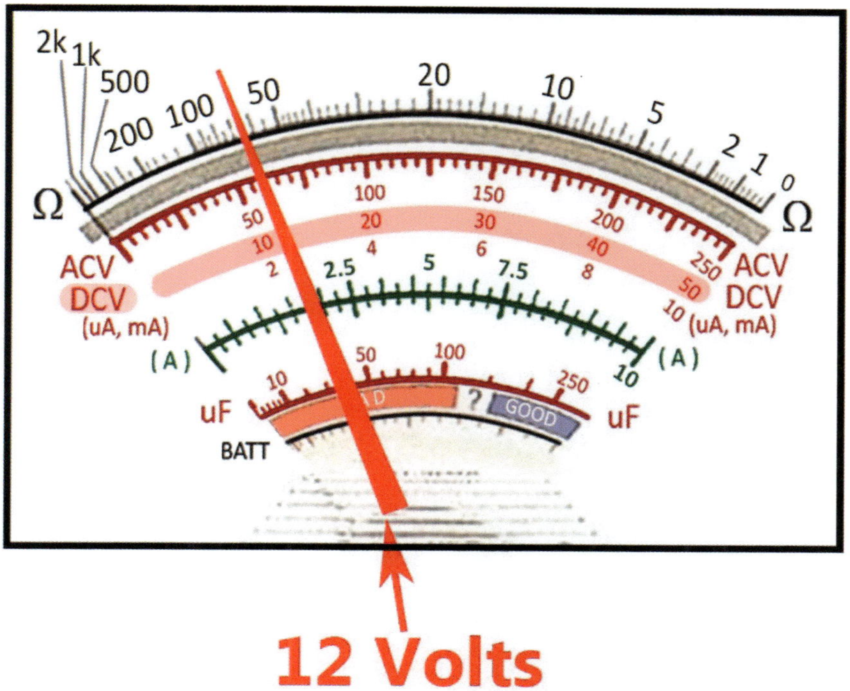

(Here we see what the meter displayed during use. The unit of measure selected was the DC Volts. The range was set to the 50 volt range setting since the battery is rated at 12 volts. The set of numbers to pay attention to are in the red number line with a maximum of 50. In this example, the number line has been highlighted for better explanation.)

In this case, the correct voltage is available at the power source. The voltage check confirms there is no problem with the charge of this battery.

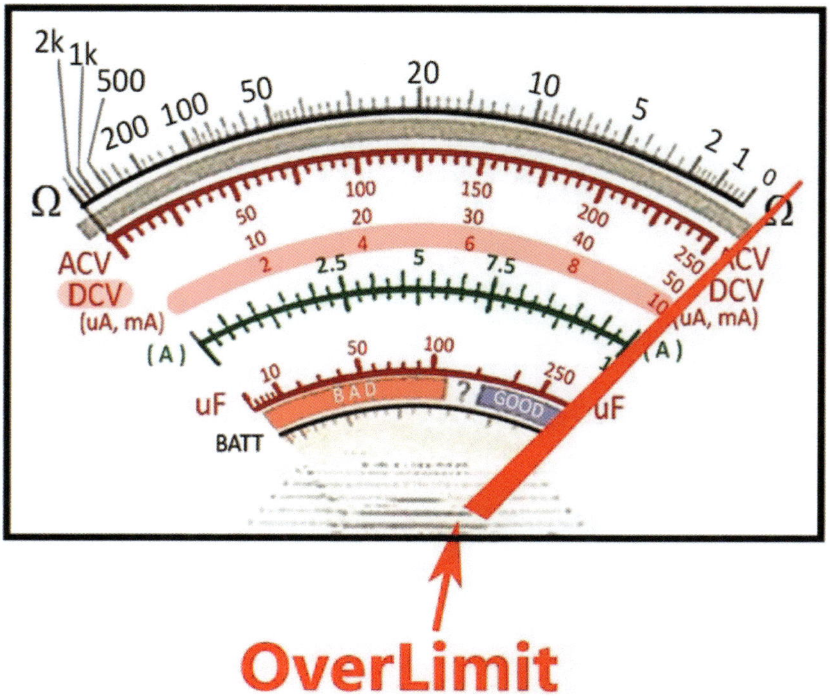

(This image shows what the meter screen will display during testing when it is setup with too low of a range setting. The 10 volt range has a max of 10 volts so the needle moved off the scale. Move the meter dial setting up one range for better resolution.)

One thing to keep in mind here is that the 10 volt range setting would have been inappropriate to use since it is too low of a range for this level of voltage. If we had selected the 10 volt range the needle would have displayed the maximum value or over limit. The following illustration is what the display would have looked like if the range selected on the meter was too low...

Now that we understand how to properly set up an analog meter for testing let us explore the other types of multimeter and the differences between them.

Manual Ranging Multimeters:

The Manual Ranging multi-meter was the next step in technology for multimeters. It is still very similar to the analog meter. In this multi-meter, the unit and unit range is also manually selected via the dial. Depending on the unit range setting selected, the maximum measurable value is also indicated. For Example: If you are on the 200v setting under DC volts, you are only able to measure up to 200 volts DC. Any measurements above 200 volts DC cannot be read and will be displayed as "OL" over the limit. To correct this, you must select a range setting higher than the reading you expect or just continue to move the dial to a higher setting until a solid measurement is seen.

In this multi-meter, the measurement display is no longer of a needle and number scale type but rather an easy to read digital number screen.

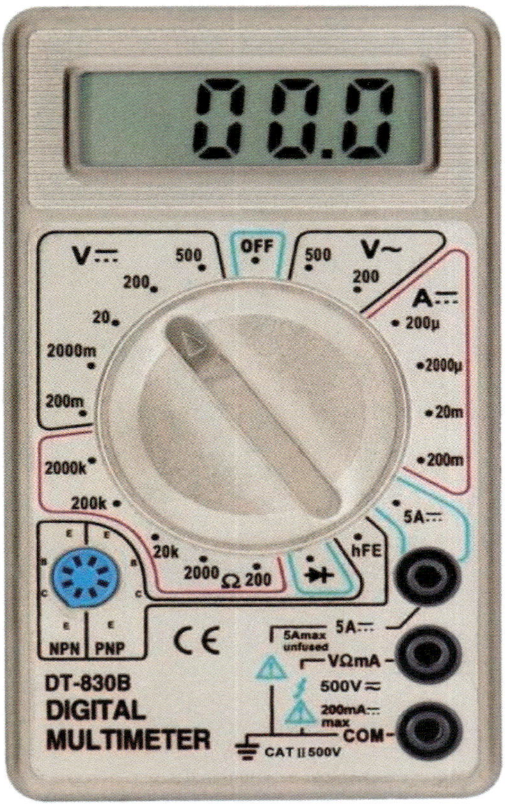

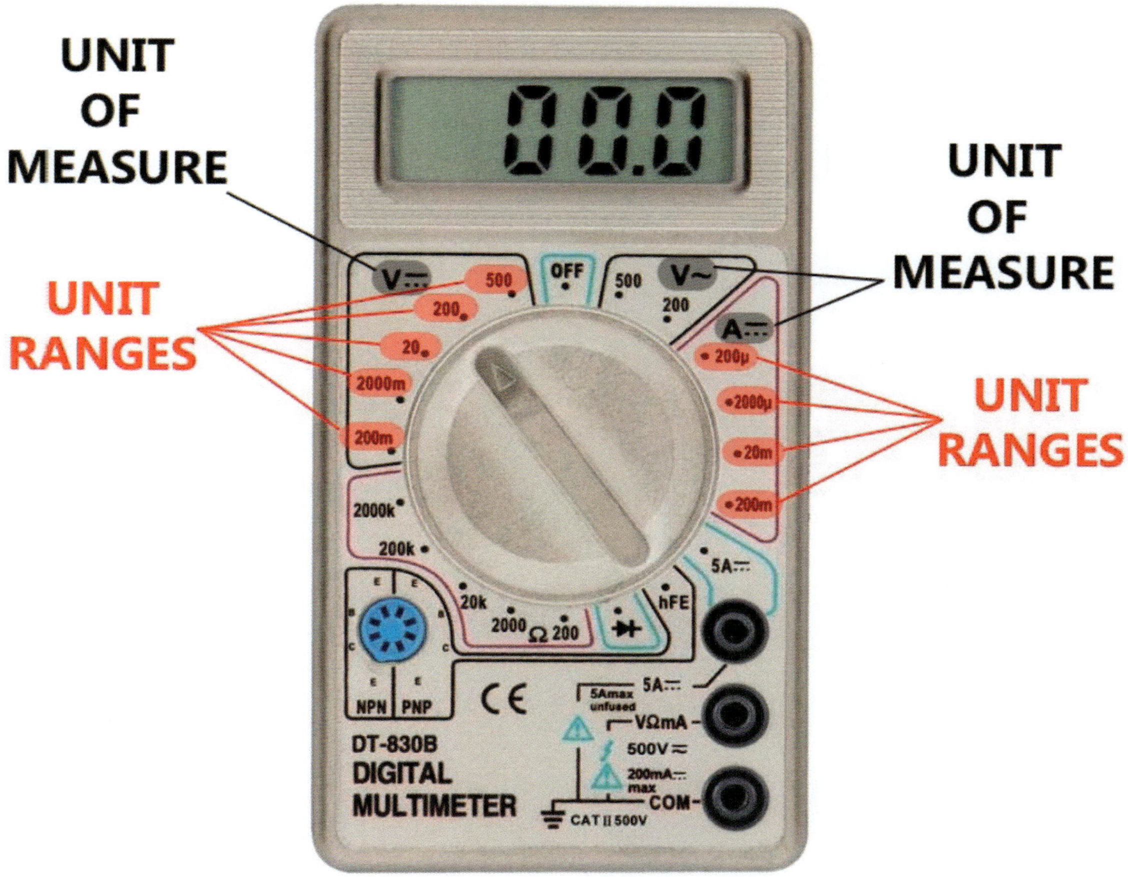

(In this picture, we see an example of a Manual Ranging multi-meter. The setting and specific range is selected through the dial of the meter. The measurements on this meter is displayed as a number value on a digital screen. As you can see around the dial of this meter, there are many unit settings to choose from. Some of the units of measure and unit ranges have been highlighted for easy identification.)

The capital "V" with the two straight lines is a unit called direct volts or DC volts and has many different ranges under it. It includes the 200 millivolt, 2000 millivolt, 20 volt, 200 volt and 500 volt ranges. All these ranges belong under one unit of measure, the DC volt. Another example is the unit "A" with the two straight lines which is known as the DC amps. This unit has dial ranges that include the 200uA, 2000uA, 20mA and 200mA. All these ranges belong under one unit of measure known as the "Amp".

Reminder About Ranges:

Just as with the analog meter, unit ranges are defined as the maximum number value that you can measure when having the meter dial in that unit range setting. In the 20mA range setting the meter will only measure up to 20mA or 20 milliamps.

If you own a similar meter that requires you to manually select ranges, start at the highest setting first. If while measuring, the number value measured on the digital display is too small in resolution, move down one range for a more accurate defined view of the measurement value.

Don't Forget the Test Probes:

One thing that will stay consistent with all three multimeters that exist so far is the need to place the test probes in the right socket.

Again, the black probe will always be installed in the COM or common socket of the meter, regardless of what dial setting you are on. The red probe will have to be installed in the socket that matches the symbol you have selected with the dial on your meter. If the dial is on an Amperage setting, you will install the red probe into the socket that has the same Amperage symbol as the one on the dial setting you selected.

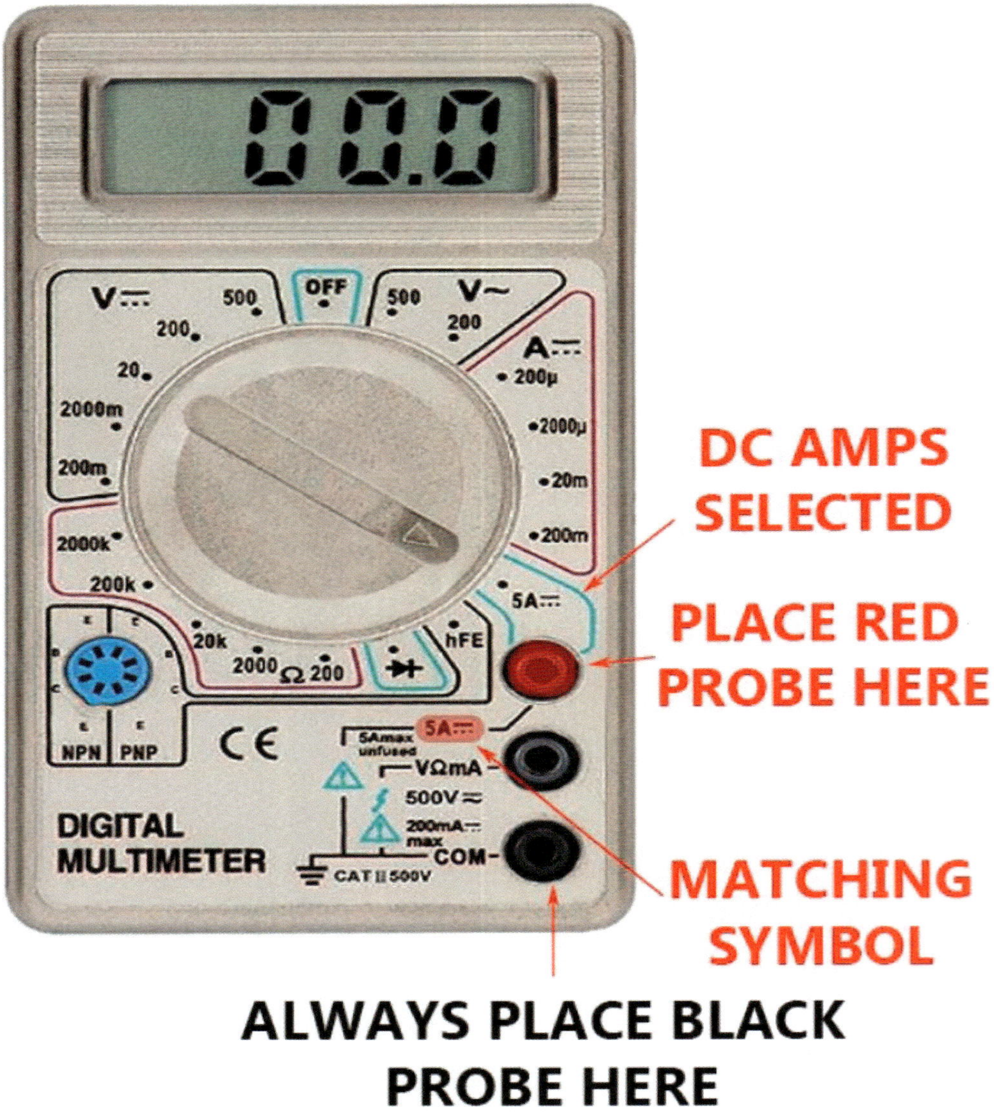

(in this example, we see the DC amp setting selected on the dial. The red probe must be installed in the meter socket that is labeled with the symbol for "Amps". The black probe always goes to the COM socket.)

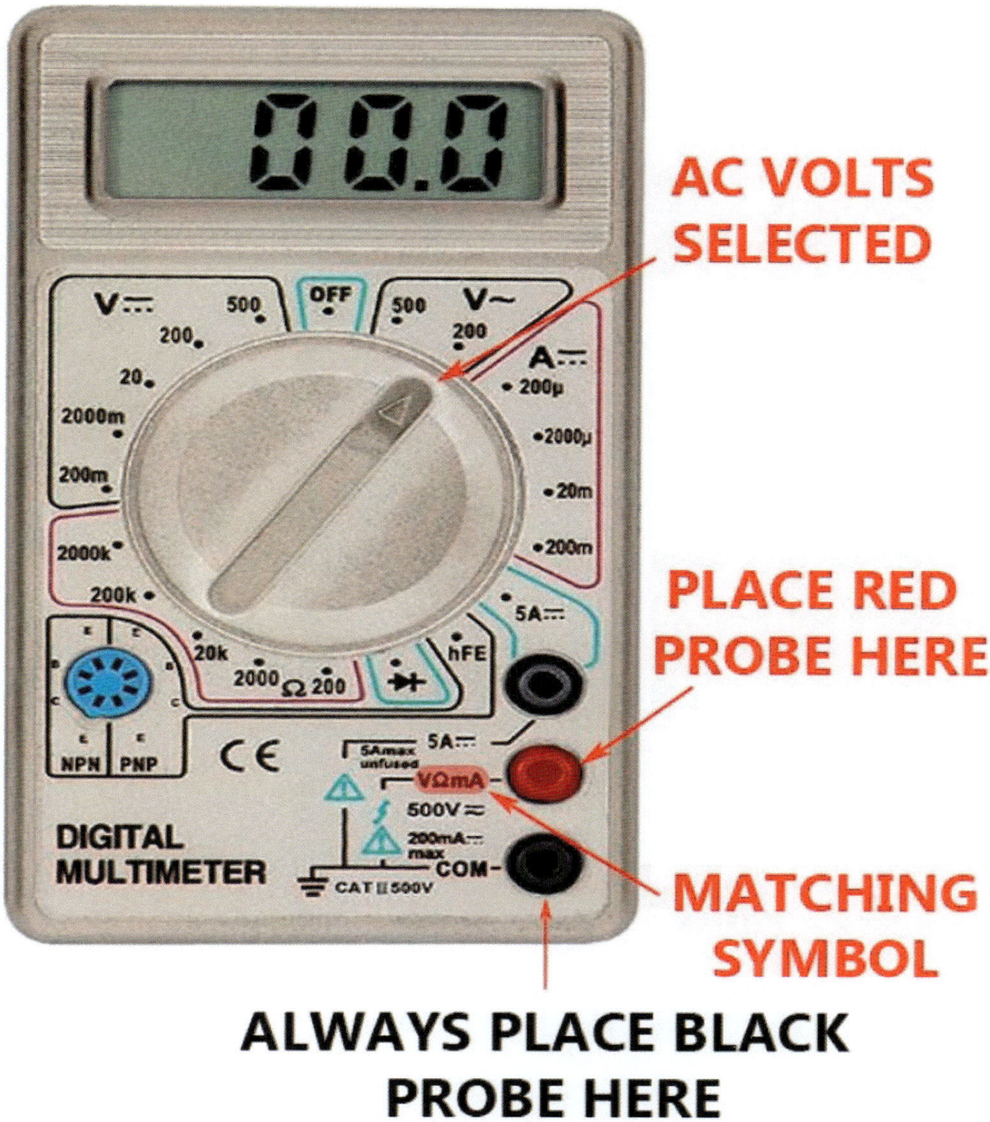

(Here we see the AC volt setting selected on the dial. The red probe must be installed in the meter socket that is labeled for Volts.)

Now that the meter is in the selected unit and unit range setting and the test probes are properly installed we can begin a measurement.

(Here is an example of the digital manual range meter being used in place of the analog multimeter on the same 12 volt battery. As you can see, it is almost identical to the analog meter except the display is a digital screen. In this case, the meter reads 12 volts indicating once again that this power source has the correct charge that it is rated for.)

Now that we understand the setup on the manual ranging meter lets take a look at the newest addition to the multimeter family.

The Auto Ranging Multimeter:

The Auto Ranging multi-meter is the most recent design and is a lot easier to use than both the previous meters. In this multi-meter, everything is simplified and the unit ranges for a unit of measurement are automatically adjusted internally by the meter. There is not need to manually select the range with this meter. All you have to do is select your unit of measure via the dial and this multimeter will automatically take care of any range issues. For Example: If you are on the DC Volt setting, regardless of whether your measurement was taken in volts or millivolts, the meter will automatically adjust to the reading that it measured and display in the best range. There is no need to select a range setting higher than the reading you expect. With this meter you can focus more on taking the actual measurement and less time on the setup.

The Meter Screen:

The display on the auto ranging multimeter will show you your measurement, along with an icon that represents the range and/or unit setting you are currently on.

(In this picture, you see an example of an Auto Ranging multi-meter. The settings are selected through the dial of the meter and the range for each setting is automatically adjusted by the meter itself. The meter screen also displays a number measurement as well as the symbol for the currently selected unit.)

Once you have set this meter to the desired unit setting and the test probes installed in the correct socket, it is ready for testing.

The Buttons and Their Functions:

RANGE: As a manual option on this type of meter, the RANGE button can be used when you are in a dial setting that can measure the unit selected in more than one range of measurement. Pressing the range button will change the range of the readings taken and will display the reading in another range.

Example 1: Milliamps (mA), Microamps (uA) and Amps (A) are the same unit, they are all a measure of amperages or Amps. The only difference is the amount or displayed range of the measured unit. Milli means "one thousandth" of something. Micro means "one millionth" of something. This means 1,000mA= 1A and 1,000,000uA=1A also. It is just a different way of reading the same unit. By pressing the RANGE button, you will see the taken measurement in another range. This is needed because sometimes in electricity, you WILL have to read very tiny values.

As a comparison, one dozen eggs or 12 eggs is the same thing, just measured differently. Sometimes you may want to measure in dozens and sometimes you may need to measure in individual eggs. The truth is that sometimes you really will have to measure in milli units or micro units because the readings you may take from electronic devices will actually be that tiny in value!

Example 2: If you wanted to measure something in millivolts(mV) but had a measurement displaying in volts(V), you would push the RANGE button until the display reads in millivolts(mV). It is the same exact measurement reading only now it is being displaying it in a different way for better resolution.

SELECT: The SELECT is used when you are in a dial setting that has more than one function or symbol on it. For example, if you were in the dial setting that contained the diode, capacitance, continuity and resistance tester all in one setting, you would use this SELECT button to switch between them to the one you desire.

***C/*F:** This button simply changes the value on the screen between measuring in Celsius (*C) or Fahrenheit (*F) when the meter is being used on the Temperature setting.

HOLD: The HOLD button is used to hold or "lock" a live measurement onto the screen of the multi-meter. Removing the probes from the test area or test piece will not affect the held reading. To reset back to normal and start a new measurement, push the hold button again.

Hz/Duty: This button simply changes between measuring in Frequency (Hz) or Duty Cycle (Duty) when the meter is on the Hz/Duty setting. I will explain more on all these in depth later once we begin testing.

(This image shows the auto-ranging multimeter version of our battery voltage check. This meter will display the measured number value as well as the current unit selected.)

Now that we have seen how the three meters function and learned how to set each of them up for testing, we can move on to learning about the basic electrical circuit that we will be using for all our testing.

Introduction To The Basic Electric Circuit:

A basic **electrical circuit** is the combination of a power source, electrical wiring, a fuse, a switch and an electrical device, connected in a way that makes the device work. In order for any electrical device to work it must first be connected to a complete electrical circuit.

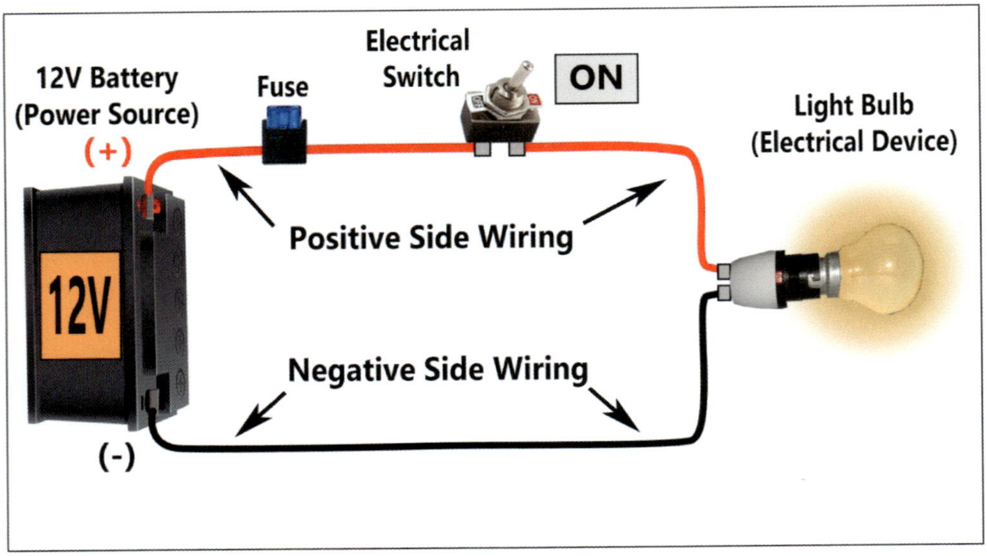

(This example illustrates what a basic electric circuit needs to include in order for an electrical device to be powered properly and safely.)

A circuit always starts off with a power source, which provides the energy required for the electrical device to turn ON. A fuse, used mainly for circuit protection, to prevent any damage to a circuit against power surges or electrical shorts. An electric switch, for user control of when the device is either ON or OFF. The electrical device or **load of the circuit** which in this case is a light bulb. And the electrical wiring that connects all the pieces together and carries the electrical energy from the power source all the way to the electrical device.

The wiring that connects to the positive side of the power source is known as the **positive or power** wiring. The wiring that leads to the negative side of the circuit's power source is known as the negative wiring. There is no physical difference between the positive wire and the negative wire of a circuit other than possibly the color of the insulation. Both are a length of wire that interconnects the electric circuit with all its part. The electrical wire is also responsible for carrying the energy available at the power source all the way to the **electrical load**. Whether the load in the circuit is a light bulb, a fan, an electric motor, a heater or any other type of device, the circuit will basically remain the same.

Note: An electrical device of a circuit is also commonly referred to as the electrical load of the circuit.

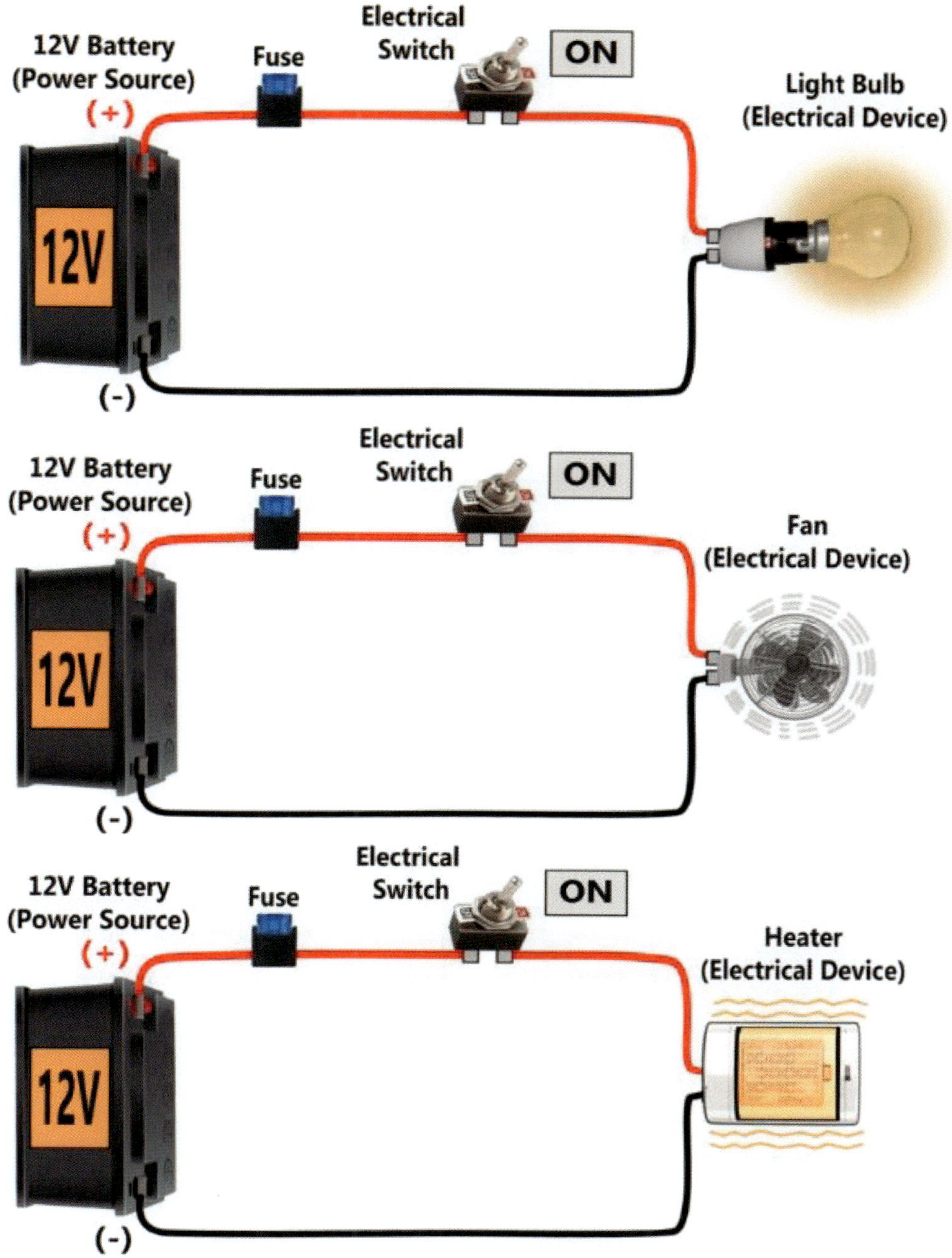

(In this image we see three examples of the basic circuit. As you can see, the basic parts that make up a circuit are almost identical. The only difference with each example is the electrical load of the circuit.)

There may be a few differences in the design of a circuit where it might require additional electrical parts but the basic essential parts of a circuit as shown in the previous image will likely remain the same as their purpose will always be needed.

In the illustrations we can see how when all the parts in the circuit are working as designed and connected properly, the circuit will provide the electricity to make the device turn ON. It is when the parts of a circuit begin to fail, become worn out or broken that the electrical problems begin.

For this book we will explore how to use each of the functions on the multimeter as well as how to understand the readings taken by the meter and gather evidence as to what is really going on inside the circuit. Now that we have seen the basic construction of an electric circuit we are ready to start using each of the multimeter functions in an actual test. To reinforce your familiarity on the the definitions of the electrical symbols on a meter please review the following chart once again.

SYMBOLS	DEFINITIONS
DCV, DC, V⎓, mV⎓, uV⎓	-DIRECT VOLTAGES (DC VOLTS)
ACV, AC, V~, mV~, uV~	-ALTERNATING VOLTAGES (AC VOLTS)
Ω	-RESISTANCE, OHMS
•)))	-CONTINUITY TESTER
—▶|—	-DIODE TESTER
—)|—	-CAPACITANCE TESTER
A⎓, DCA, mA⎓, uA⎓	-DIRECT AMPERAGE (DC) CURRENT FLOW (DC) AMPS
A~, ACA, mA~, uA~	-ALTERNATING AMPERAGE (AC) CURRENT FLOW (AC) AMPS
Hz, FREQ	-FREQUENCY TESTER (HERTZ)
DUTY CYCLE	-DUTY CYCLE TEST
TEMP, C, F, K	-TEMPERATURE MEASURE

(This chart reviews all the common symbols or abbreviations you will see on a multimeter.)

You may find that there are more symbols on your meter that I may not have included in the chart. The ones I have included are the only ones you will ever use professionally to solve a problem. Keep these definitions in mind and handy when selecting your measurement setting with the dial on your meter.

In the next chapters, I will be showing you examples of meter testing using the digital auto-ranging multimeter simply because it is more common and easier to understand. Just keep in mind that the knowledge of how to set up the other types of meters will allow you to use any multimeter for our testing.

Ch.2: Voltage AC and DC

In the following sections we will review all of the important concepts to remember before beginning voltage testing.

What is Voltage:

Voltage, or volts, is the word used to describe the measurable unit for **electrical pressure.** This electrical pressure (or voltage) is one of the three main units of electricity required for an electric circuit to work properly.

A voltage measurement is a measure of the amount of electrical pressure inside of an electric circuit. This voltage (or pressure) inside the circuit is what is responsible for actually pushing electricity throughout the circuit for it to work. Without voltage, there can be no electricity flowing.

Just like how a water pipe circuit in your household needs water pressure, provided by the water company's pumps, to make the water in the pipes flow all the way up to your house. The water pressure from the pump is what makes water move throughout the pipes. Likewise, voltage from an electrical power source is what creates the flow of electricity.

How Voltage Creates The Flow Of Electricity:

For electricity to start to flow, there must exist inside of a circuit a high voltage (high pressure) and a low voltage (low pressure). The two pressures must then be exposed to each other or be connected to each other in some way for the process to begin. When you expose a high pressure to a low pressure the higher pressure will always try to move towards the lower pressure. For example, when you expose a gas with a high pressure to a lower pressured gas, the higher pressured gas will try to move towards the lower pressure gas.

This is one of the basic laws of pressure which is normally learned as part of a physics class. The reason why it is important in electricity is because it allows you to understand how things work. This is a law that also applies to all other kinds of pressures including voltage. A second example of this law of physics involves gas pressure from a regular soda bottle. The gas inside the soda bottle is a higher pressure than the outside air pressure. When the bottle is opened the higher gas pressure inside the bottle is exposed to the lower gas pressure outside the bottle. The higher gas pressure then tries to move out of the bottle toward the lower pressure area until the two are the same pressure and no difference is pressure exists anymore.

To not complicate the explanation too much on how things work, the main takeaway is that when two different pressure levels are exposed to each other, the higher pressure will move towards the lower pressure. In the case of an electric circuit, the higher voltage will move towards the lower voltage. In the process of the voltage moving, it will push the electricity forward and force it to flow throughout the circuit.

In a complete circuit, the positive side wire of the battery (high voltage) is connected to negative side wire of the battery (low voltage) with an electrical device in between the two wires. When you put a device in between the two different pressures, you can actually take advantage of this law of physics and use the energy from the high pressure moving towards the low pressure to create something useful.

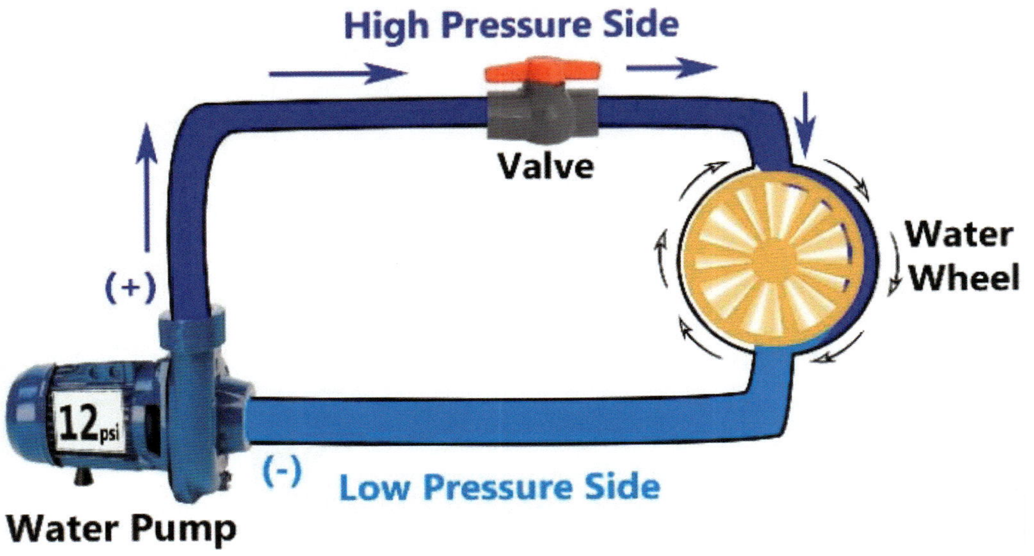

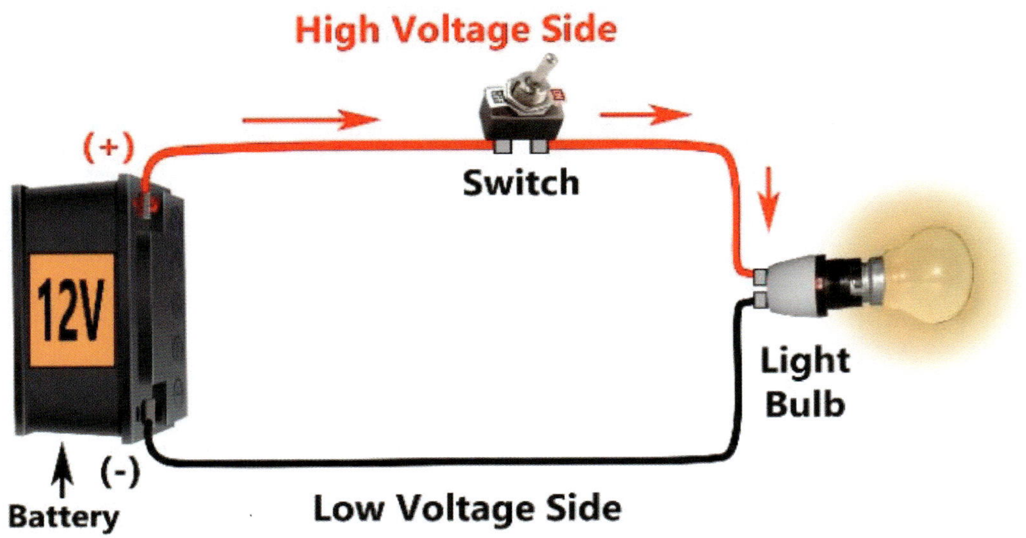

(This image represents the similarities between water flow and electrical flow. When you have a high pressure and a low pressure connected to each other with a device in between, you can make that device work. The high pressure water coming out of the pump will move towards the low pressure water and in the process make the water wheel move. In the case of the electric circuit, the high voltage electricity will move towards the low voltage side and in the process turn on the light bulb. This process will continue until either the pump or the battery stop providing the higher pressure or the circuit is turned OFF.)

Note: This rule applies to fluid pressure, electrical pressure, gas pressure or any other type of pressure.

Notice that when the pressure reaches the device in a circuit the pressure comes out lower on the other side of the device. The reason why is because the energy from the high pressure is actually used up in order for the device to work. When the high pressured water reaches the water wheel, the energy will be used up by the water wheel in order to make the wheel spin. The water exiting the water wheel will have a lower pressure for this reason. Similarly, in an electric circuit, the energy from the high voltage electricity trying to move towards the low side is converted at the electrical device into another kind of energy, this case light energy. As mentioned before, this process will continue until either the pump or the battery stop providing the higher pressure or the circuit is turned OFF preventing the high pressure from reacting to the lower pressure.

By placing something in the way of the moving pressure you can harness the energy to create a working circuit. This is the basic idea of how we use one of the laws of physics to purposely make electricity flow by taking advantage of how pressures behave when exposed to one another.

Completing The Circuit

For the circuit to work, there must always be a complete connection or path between the positive side and the negative side. If the connection is ever broken, the high voltage will no longer be exposed to the low voltage and movement of the electricity will stop.

With the circuit turned ON, voltage must always be present up to the electrical device in the circuit. This voltage at the device can be measured to confirm that there is nothing wrong with the circuit. If there is ever a situation where voltage is missing or measured by the voltmeter somewhere where there shouldnt be any voltage, this is evidence that an electrical problem exists. Depending on where in the circuit the voltage is measured, this can lead us to a conclusion of what is actually happening.

System Voltage of a Circuit:

The amount of voltage or pressure needed for each circuit is not always the same. Each electrical circuit is engineered to work with a specific voltage in order for it to work correctly. The **system voltage** for the circuit will depend on the amount of voltage that the power source installed in the circuit is able to provide. Different circuits may require different power sources, therefore needing different levels of voltages for it to work. This can be compared to how there are different water pumps available that are rated to be able to produce different levels of water pressures needed for the different types of water systems that there are. Similarly, there are many electrical power sources available that can produce different levels of voltage.

If a circuit is rated to work on 28 volts, all the electrical parts are specifically designed to work correctly with that level of voltage. If you provide less or more voltage than the circuit is rated at, it will not work properly or not at all.

The Types of Power Sources:

There are two main type of power sources in the electrical world. Examples of power sources include batteries, generators, solar panels, capacitors, wind mills, etc. Basically, any device that creates electricity for a circuit to work is an electrical power source. The two types of power sources to know about are either AC or DC power sources. The type that is used in the circuit will determine which of the two volt settings to use on the dial of your meter. An **AC power source**, such as an AC generator, produces a very different kind of electrical pressure (voltage) than a **DC power source**, such as a battery.

AC Power Sources:

The **AC voltage** setting on your meter will be used when measuring voltages from circuits powered by an AC power source. This includes household power outlets, household wiring, Industrial wiring and/or practically any other electric circuit powered by an AC generator.

A light bulb that is rated to work on 120 volts AC requires that it be powered by an AC power source (ex. 120 volt AC generator) that is rated to provide 120 volts AC continuously for the circuit. This voltage can be observed and measured with your AC voltmeter when the you place the probes across the electrical device for the circuit.

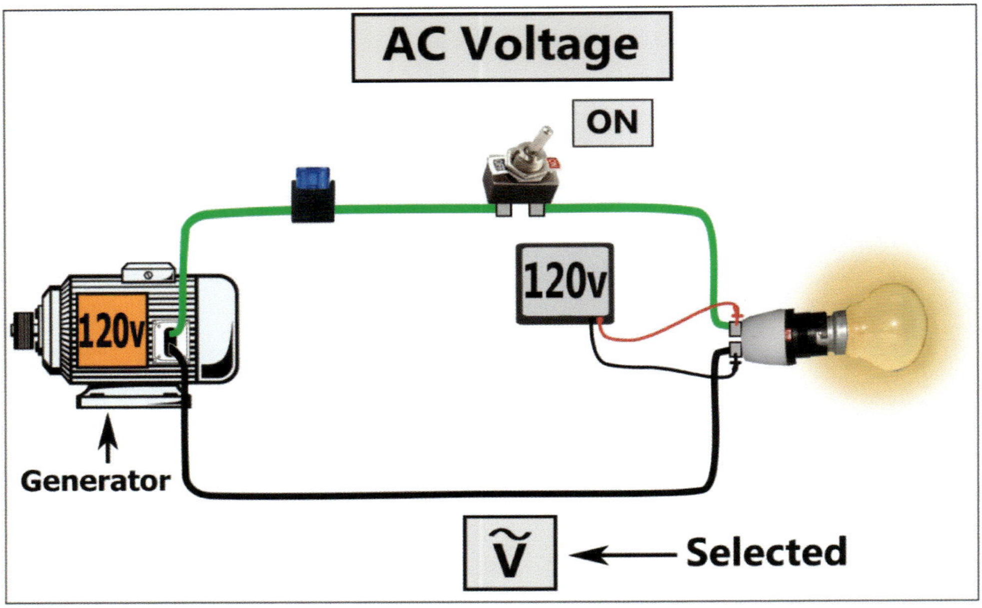

(This illustration shows a 120 volt AC circuit and the AC voltmeter setting selected for use. In this image, the probes of our AC voltmeter are on each side of the light bulb. You also need to have the circuit ON to see what voltage is available at this AC powered light.)

DC Power Sources:

The **DC voltage** setting will be used when measuring voltages from circuits powered by a **DC power source**. This includes many automotive, motorcycle, aviation, some industrial electric circuits or practically any other circuit that is powered through a battery, capacitor, a solar panel or a DC generator.

A light bulb that is rated to work on 12 volts DC requires that it be powered by an DC power source (ex. 12 volt Battery) that is rated to provide 12 volts DC continuously for the circuit. This voltage can be observed and measured with your DC voltmeter when the you place the probes across the electrical device for the circuit.

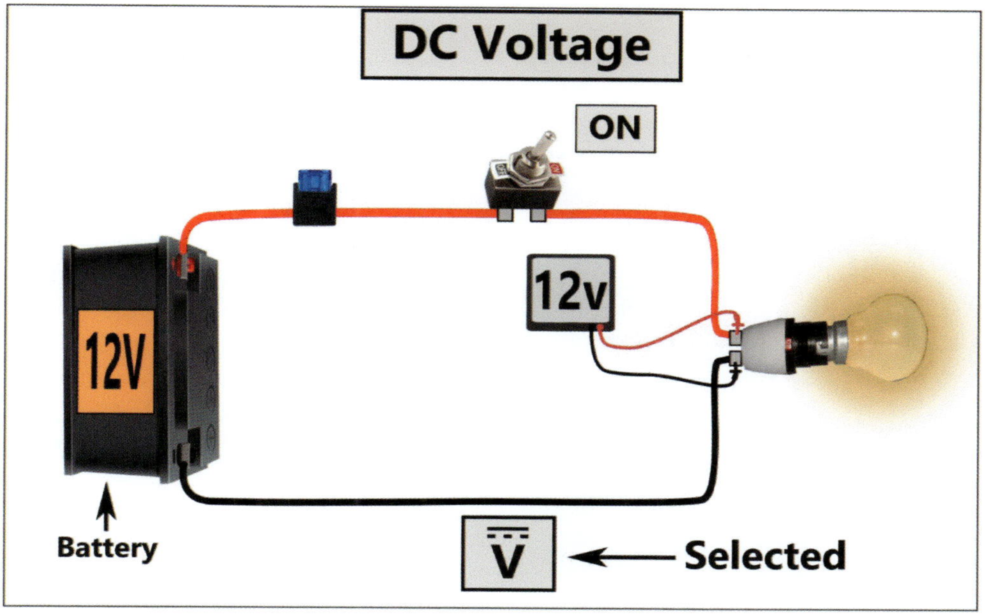

(This illustration shows a 12 volt DC circuit and the DC voltmeter setting selected for use. The test probes of our DC voltmeter are connected to each of the two terminals of the light bulb. For this test to work, you also need to have the circuit ON to see what voltage is available at the light.)

In either of these two voltage settings you are dealing with measuring the voltages or the electrical pressure that makes the circuit work. Remember that every circuit needs voltage to make electricity flow inside the circuit, just like how the water in the water pipe circuit needs pressure from a pump to make it flow.

You can test for voltage at the electrical device to confirm that all the voltage is available at the **load** of the circuit. Whether it is AC or DC voltage, the hook up to a test area is the same. Just keep in mind that electrical parts designed to work on AC voltage will not work on DC circuits and vice versa.

How the Voltmeter Actually Measures Voltage:

Contrary to what most beginning electrical technicians assume, the voltmeter does NOT directly read voltage. The voltmeter actually reads **the difference in voltage between two test points.** THIS is very important to keep in mind. The voltmeter reads **Voltage Difference** (in other words pressure difference) between the two test probes. It works by sampling the voltages at the locations of each of the two test probes. It takes each sample of the voltage that is present where you placed each test probe and then subtracts the values and displays the subtracted answer on the meter screen.

For example, if you have one test probe attached to somewhere in a circuit that has 12 volts and the other test probe connected to a part in the circuit that also has 12v, the meter will display 0 volts. 12v-12v=0v.

The reason why this is important to know is because by knowing how the voltmeter works you will understand how you can use it to gather evidence of what is happening inside a circuit.

Voltage Levels Throughout A Circuit:

In the following examples we will begin our voltage testing while also taking a look at what a normally working circuit reads in terms of where voltage SHOULD be present in the circuit.

Checking For Proper Voltage Available at The Power Source:

It is one of the jobs of the voltmeter (multimeter on the voltage setting) to measure and confirm that voltage coming from the power source of the circuit is the correct level and present all the way up to the electrical device that the circuit is supposed to be powering. The voltage at the power source must be correct when the circuit is turned ON and the power source is actually being worked.

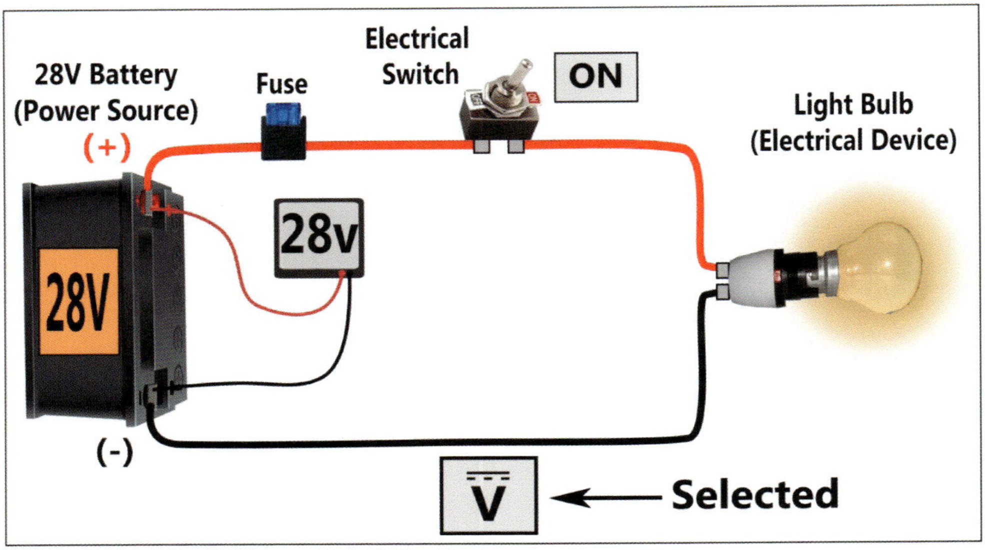

(This example shows an image of a 28 volt circuit being tested by the voltmeter to confirm that the voltage that is present at the power source is actually the correct level with the circuit turned ON. One test probe to the positive side of the battery, one test probe to the negative side of the battery. The multimeter is set to the symbol that represents DC volts. The meter displays a reading of 28 volts indicating that the power source is working and is providing the correct voltage for the circuit to work.)

As shown in the previous example, if you place one of the meter's test probes on the negative side of a battery (which has 0 volts) and then the other probe on the power side of the battery that had (28 volts), then the voltmeter will display the difference in the pressures. In this case 28v-0v is 28 volts. The multimeter reading of 28 volts indicates that the correct voltage is being provided when the battery is under load.

Note: Whenever you have one test probe attached to the negative (0 volt) side of the power source, you can use the other test probe to check for voltages across various places in the circuit.

Testing a Power Wire:
With the voltmeter it is also possible to probe the power wire of an electric circuit to check to see if there is the correct voltage throughout the entire power wiring. With the circuit turned ON, the power wire SHOULD have the voltage from the power source available at any point in the wiring up to the electrical device that the circuit is powering.

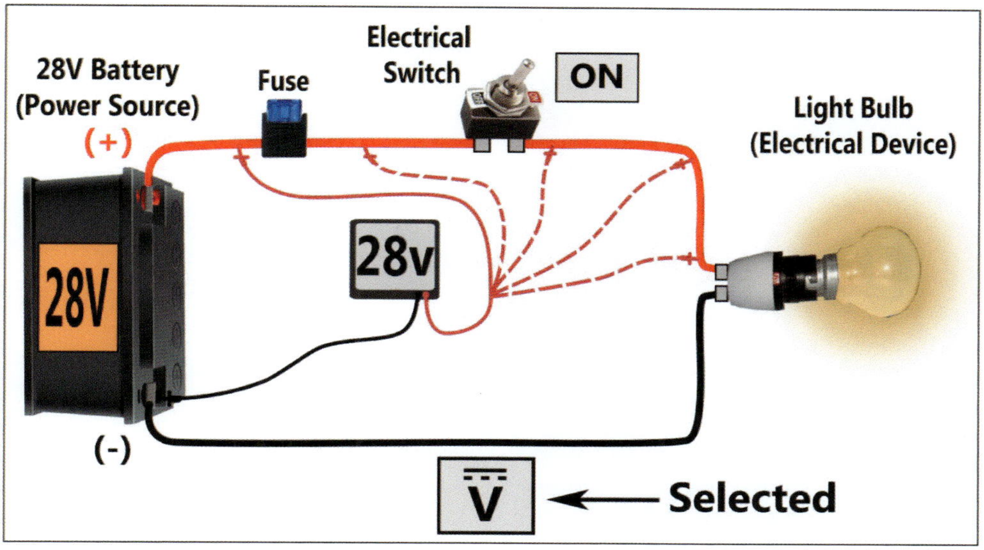

(This image shows how in a working circuit, the power side should have the voltage from the power source (this case 28 volts) available at any given point in the power side wiring. The voltmeter is set to DC Volts. One test probe to the 0v negative side of the battery and the other test probe being used to test various locations on the power side of the circuit.)

This voltage testing is performed to show where voltage SHOULD be and also to make sure that the correct voltage is present all the way up to the electrical device. If it is present at the device, this means that the power wiring and all the other parts on the power side of the circuit are good. If there was ever a a big enough loss of voltage somewhere in the power side of the circuit, the light bulb would not light and a problem would be indicated.

With this test setup, if you found a **zero** voltage reading somewhere in the power side where there SHOULD be 28 volts this is proof of a problem. The problem may be caused by the wiring or another part on the power side of the circuit that may be broken somewhere. This would be considered an **Power Side Open Circuit** and would cause the light to not turn ON. Further diagnosing of the circuit would be necessary to find this problem.

Checking For Available Voltage at The Electrical Device.
The next step in our walkthrough across a properly working circuit is to check how much voltage actually made it to the electrical device. Ideally you would want all the voltage that was available at the power source to be available at the electrical device when the circuit is turned ON.

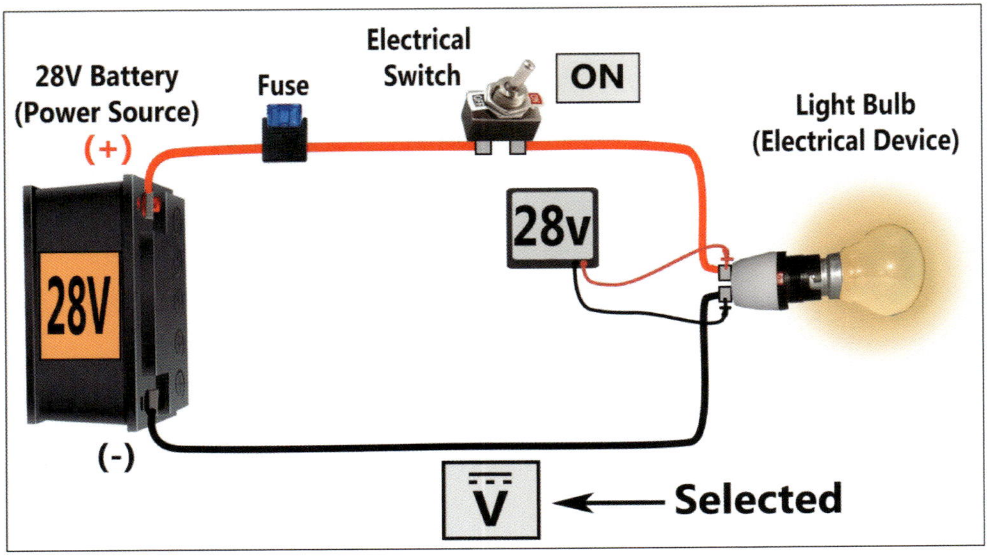

(Image of a 28 volt circuit being tested to confirm that proper voltage from the power source is available up to the electrical device. One test probe to the positive side of the light bulb, one test probe to the negative side of the light bulb. The test requires that the electrical switch for the circuit be in the ON position and the multimeter set to the symbol that represents DC volts. In this example, we have used the terminals of the light bulb as test points for our test probes.)

Again, the test probes are placed somewhere where the difference in voltage between the two probe is 28v. Red probe to a point in the circuit with 28v and the black probe at a point with 0v. 28v-0v= 28v.

Note: With the circuit ON, the negative side of the circuit should ideally have 0 volts. The positive side of the circuit should have all the voltage that the power source is able to provide all the way up to the electrical device. The reason for the 0 volts on the negative side is because once the voltage reaches the electrical device, the voltage is used up by the device to create a certain effect. If the device is a light bulb, the voltage will be used up and converted to create light. If the device is a fan, the voltage will be used up to make the motor spin the fan blades. If the device were a heater, the voltage would be used up and converted into heat.

In the illustration we see the voltmeter performing a quick check for voltage at the electrical device. With the switch turned to the ON position, the voltage for the circuit should always be available up to the electrical device.

If there is an electrical problem where the device not turning on and voltage was confirmed by the voltmeter to be available at the device, then the problem is not something in the rest of electric circuit. This voltage test would be proof that the problem is either a bad connection at the electrical device or a bad device itself, this case the light bulb.

As a second possibility, if there was an electrical problem where the voltage was NOT available up to the electrical device, then this is proof that the problem is somewhere in the other parts of the circuit and not the device.

Quick Tip: By checking for voltage at the electrical device first you can eliminate a lot of time wasted testing the circuit at every point. If the voltage is confirmed to be at the device while the circuit is ON, then this means that there is nothing wrong with the rest of the electric circuit. If the voltage wasn't there with the circuit ON, then this means your device is not the problem and the problem is something else in the circuit.

One thing to understand about voltage tests is that due to the amount of information you can gather using only the voltmeter, it will become your best tool when troubleshooting electrical problems. You will understand more as to why this is true later on, but for now let's continue exploring how the voltmeter works and how else we can use it.

Testing A Negative Wire:

Using the voltmeter also makes it possible to probe the negative wire of an electric circuit to check to see if there is the proper voltage level throughout the entire negative wiring. The negative wire SHOULD ideally have **zero** voltage available at any point in the wiring because the voltage should have been all used up by the electrical device in the circuit (this case the light bulb).

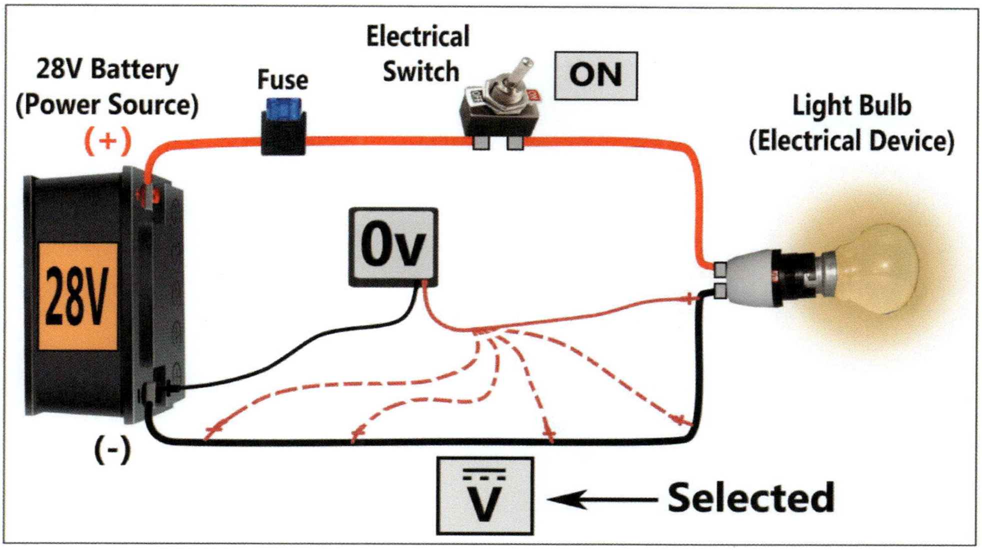

(This illustration shows how the negative wire should have zero voltage available at any given point in the wiring. Checking to see if there is the ideal zero voltage all the way up to the electrical device means that the negative wiring leading back to the battery is good.)

Both of the test probes are connect to points in the circuit where there is the same 0v available so the voltmeter will display the difference. Black probe at the battery negative (0v) and the red probe at various points on the negative side wiring. 0v-0v=0v.

If there was ever an electrical problem present that caused the power source voltage (28v) to be read somewhere in the negative wiring, it is because the negative wiring must be broken somewhere. This is considered a **Negative Side Open Circuit** and would also show itself as the light not turning ON.

With this test we conclude our walkthrough of where the voltage SHOULD and SHOULDNT be present in a properly working circuit.

Voltage Between Two Similar Voltages.

To make sure you understand the concepts of how a voltmeter works we will look at another circuit example of what the voltmeter reads when testing two similar voltages. The power source voltage for this circuit is now 12 volts. if you have one test probe on a part of the circuit that has 12 volts and then the other probe on a different test point in the circuit that also has 12 volts, then the voltmeter would display the difference in these pressures. In this case 12v-12v is 0 volts.

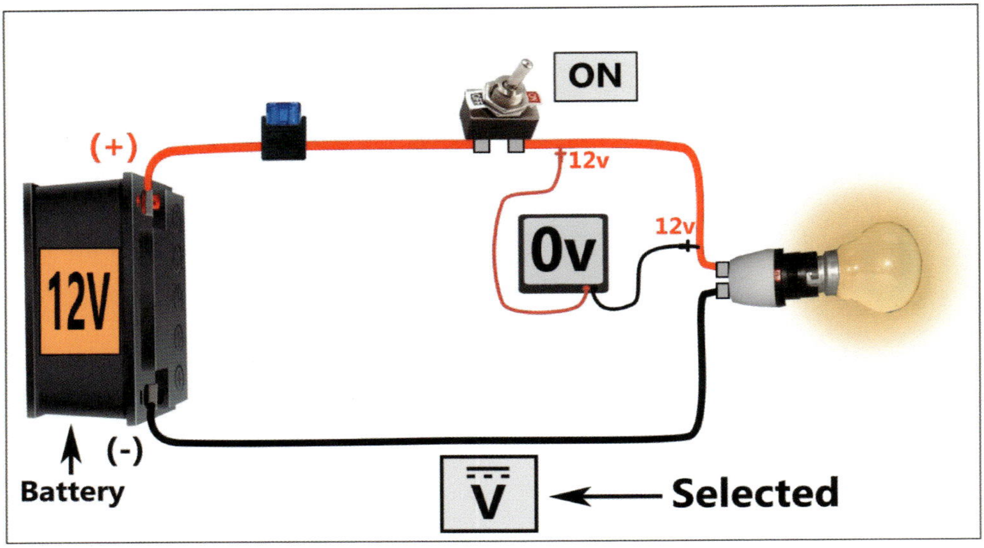

(Example of what the voltmeter will display when testing at two points that have the same voltage.)

This test reading can be very confusing because you might think there is no voltage in the circuit when in fact there is voltage. The way to check for voltage in a circuit is to place one of the test probes on the negative side of the battery (0v) and use the other probe to "look" for the voltage in other points on the circuit. This will ensure that whatever point the other test probe touches in the circuit during testing is being compared to 0 volts and is the true voltage pressure available at that specific point. Always remind yourself that the voltmeter reads **voltage difference** between where the two probes are placed.

Voltage Between Two Different Voltage Levels.

If you have one test probe on a part of a circuit that has 12 volts and then the other probe on a part of the circuit that has 4 volts, then the voltmeter screen would display the difference in the two test points. In this case 12-4v is 8 volts. For this example, we have connected two electrical devices in line (or **in series**) with each other on our basic circuit to explain this concept.

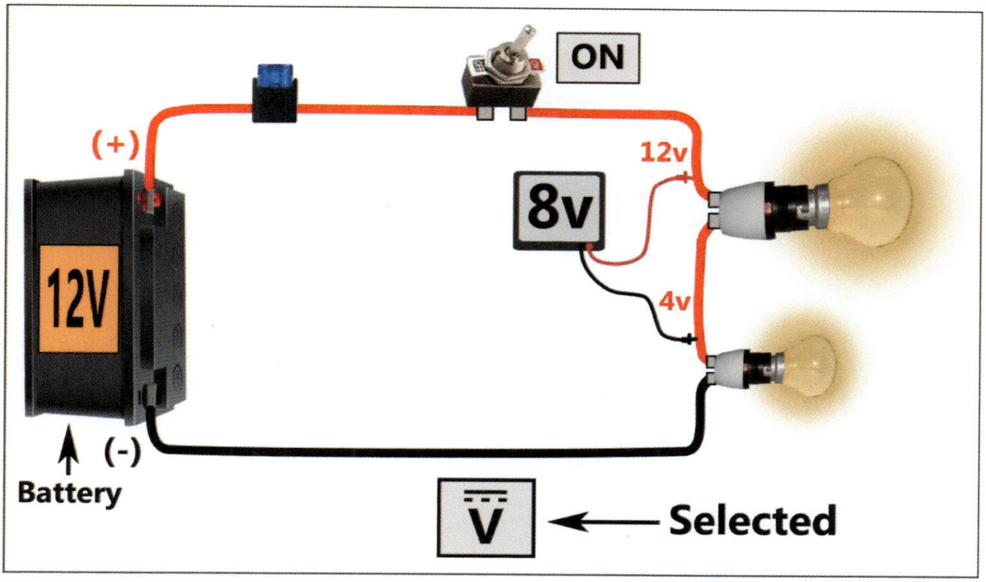

(This image shows how the voltmeter determines what to display on the screen. There are two light bulbs in this circuit that are installed inline (or **in series**) with each other. Because they are in series, each of the light bulbs will use up a certain amount of the available 12 volts. The voltmeter probes are connected across the positive side of the larger bulb and the positive side of the smaller bulb. The meter reads the voltage difference between these two test points, this case 8 volts.)

Notice that this voltage test is actually putting the meter probes across the larger bulb. Also notice that the voltage available across the large light bulb is NOT the full voltage from the battery. This is because the voltage that is available from the battery is shared between the two electrical devices when there is more than one device in series. The amount of voltage at each device in series will add up to the total available voltage, this case 12 volts. The larger light bulb has 8 volts across it and the smaller light bulb has 4 volts across it. This adds up to the total available voltage which is 12 volts in this case. This is considered normal when there is more than one electrical load.

Voltage Differences Involving the Electrical Switch:

As you may have already noticed, I have continued to remind you that the voltage tests we have done so far require the electrical switch to be in the ON position. This will be true for the majority of your voltage testing. You always want to test the circuit live and with the switch ON to get a true result from your voltage readings.

Well what if the switch is in the OFF position? Where will the voltage be present then? Let's explore the differences between when the switch is ON vs OFF.

In the OFF position, the electrical switch acts like a break in the circuit. This can be considered an intentional **open circuit,** which prevents the voltage from the battery from making it all the way up to the electrical device. It also prevents any electricity from flowing since the circuit is technically "broken" at the switch. Because there is no voltage at the device, the light bulb will not light.

It is the job of the switch in the OFF position to prevent any electricity or voltage from passing it. Because of this, you will see all the available voltage from the battery only make it up to one side of the switch. On the other side of the switch there will be 0 volts.

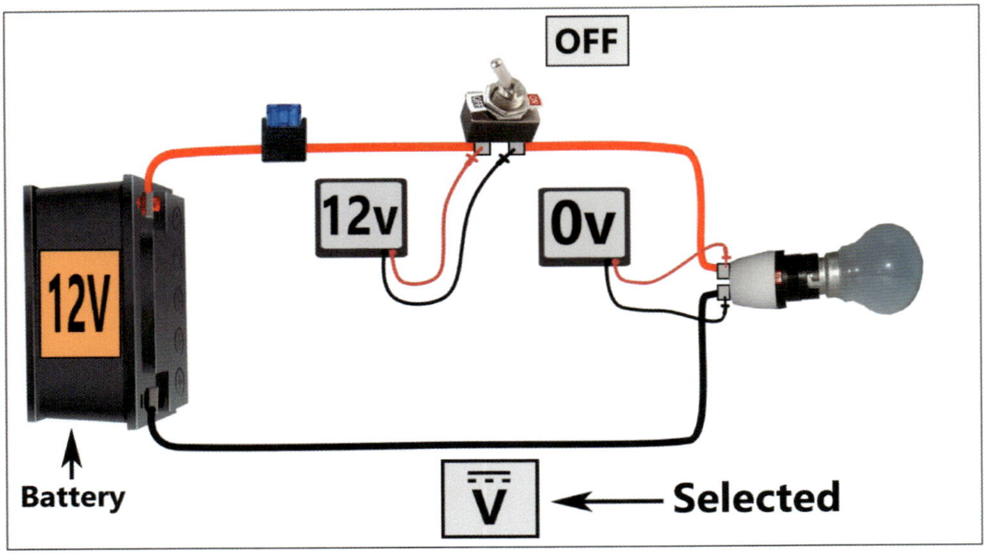

(This image represents where the voltage is present with the switch OFF. Because of the voltage difference now at the switch, putting a voltmeter across it will read full power source voltage. On the other hand, at the light bulb the voltmeter will read 0 volts.)

With the switch turned to the ON position, the voltage will move past the switch and once again travel towards the light bulb where it will be used up to create light.

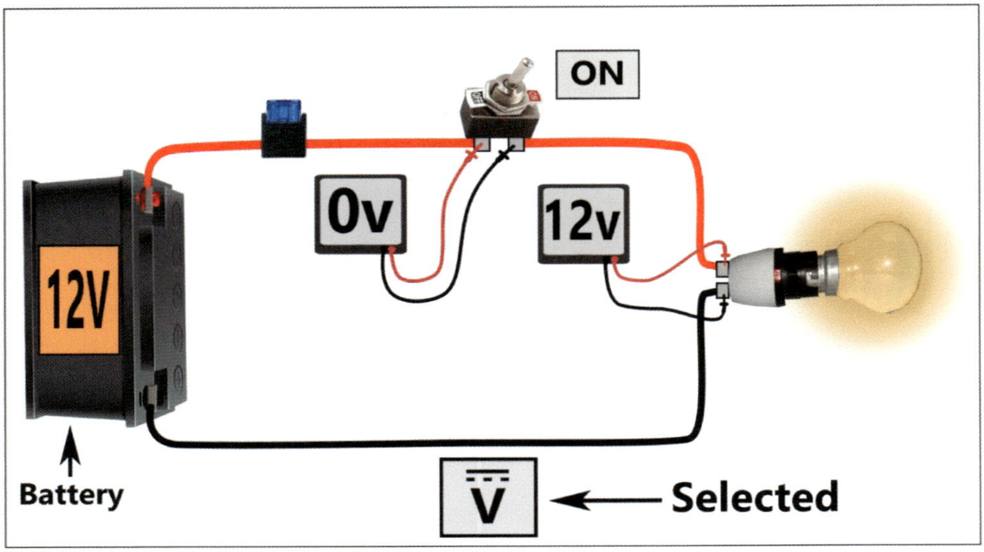

(This image represents where the voltage is present with the switch ON. Because of the voltage now passing through the switch, putting a voltmeter across it will read 0 volts. The two sides of the switch now have the same 12 volts so the voltmeter across the switch reads zero. At the light bulb, the voltmeter will now read 12 volts.)

Now we know that with the switch ON, we should have 0 volts across the switch. With the switch OFF we should have full power source voltage across it. This is actually the way to test if the switch is working. For example, if you had 0 volts across the switch regardless of the switch position, this indicates a problem with either a shorted switch or there is a lack of voltage from another problem happening before the switch. If instead you had problem where the meter reads 12 volts across the switch regardless of its position, this would indicate that the switch is internally bad and unable to complete the circuit when switched to the ON position.

With the circuit ON, there should be 12 volts present throughout all the power side all the way up to the light bulb. If there is ever a reading of 0 volts somewhere on the power side, with the switch ON, this indicates a break somewhere in the circuit.

Types of Open Circuits:

A break in the circuit or **open circuit** is when there is no longer a complete and continuous path from the positive side of the power source to the negative side of the power source. When a circuit is open, all electricity stops flowing in the circuit. As seen before, the voltage for the circuit will NOT be present at the electrical device as it should be. Instead, the voltage will only make it up to the point where the break in the circuit is located. By using the voltmeter, we are able to identify if there is an open circuit problem by testing at the electrical device with the circuit ON. If there is no voltage at the device, this may indicate an open circuit problem.

Positive Side Open Circuits:

The location of the **open** in the circuit will have the full power source voltage across it. It can be compared to an electrical switch in the OFF position. The path for electricity to flow is broken. One side of the break will have power source voltage and the other side will have 0 volts. This fact can be used to logically determine where the break in circuit is by using the voltmeter.

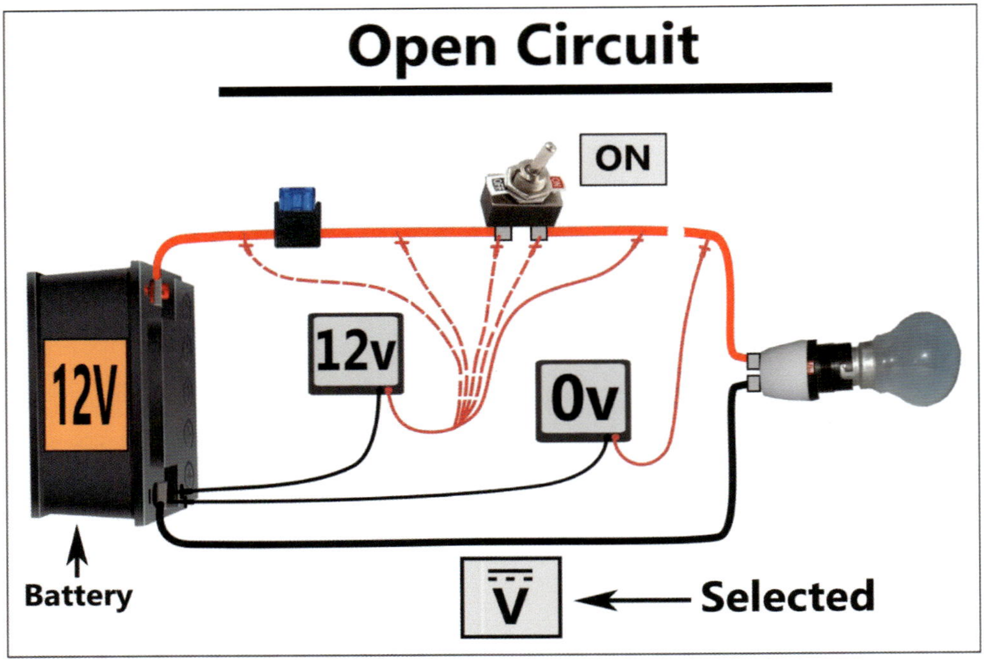

(Example of a power side open circuit being located by using the voltmeter. One test probe is connected to the negative side of the battery and the other being used to test various points in the positive side of the circuit. While testing the power side at various points, any 12v reading means that there IS voltage up to that point and the problem is still somewhere ahead in the circuit. If at some point the meter reads 0 volts on the power side, this indicates that you have just passed the location of the break. The open circuit is between the current test point and the last test point where the meter still read 12 volts.)

As shown by the voltmeter, anywhere before the break in the circuit will still read power source voltage. Anywhere after the break will read 0 volts. By understanding this you will be able to use your voltmeter to narrow in on the location of the problem and fix it.

Negative Side Open Circuits:

The same concepts from before applies to a negative side open circuit as with a positive side open circuit. All the voltage of the circuit will be seen across the break in the circuit.

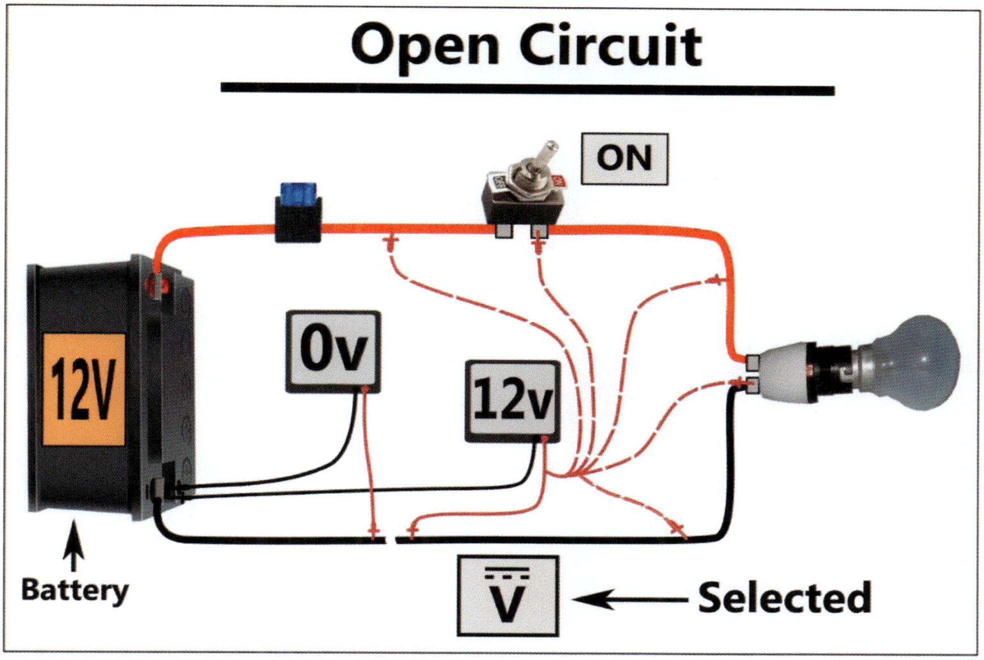

(Example of a negative side open circuit being located by using the voltmeter. One probe is connected to the negative side of the battery and the other being used to test various points of the circuit. While testing the circuit at various points, any 12v reading means that there IS still voltage up to that point and the problem is somewhere ahead in the circuit. If at some point the meter reads 0 volts, this indicates that you have moved past the location of the break. The open circuit is between the current test point and the last test point where the meter still read 12 volts.)

Regardless of whether the break is on the positive side or the negative side of the circuit, the open circuit problem will read power source voltage across it. As shown by the voltmeter, anywhere before the break in the circuit will still read source voltage. Anywhere after the break will read 0 volts. By understanding this you will be able to narrow in on the location of the problem using the voltmeter as shown in the previous images.

For either kind of open circuits, further diagnosing of the circuit would be necessary to effectively find these problems. I have introduced only the basic idea on how to identify an open circuit using your meter setting but the actual problem finding and solving can be a little more complicated and beyond the scope of the lessons in this book. The purpose of THIS book is to teach the basics of electricity and explain how each of the multimeter settings work.

If you are interested in more about solving Open Circuit problems and much more, please check out my other book “How To Test Circuits Like a Pro” where I have outlined exactly how to solve these types of electrical issues. In that book you will be shown how the circuits react to different kinds of electrical problems and a step by step training on how to solve them.

Note: When performing tests using your multi-meter, any measurements in any setting should be written down and analyzed or compared to good values to see if they are within proper specs.

Tips To Make A Good Connection:

In order for your meter to get a stable voltage measurement you must have a solid connection to the circuit that will not move on you or give you false readings. One of the biggest problems with beginning electricians is trying to figure out how to connect the probes of the meter to the circuit under test. If you don't make a solid connection to the circuit you may get a false reading that may lead you down the wrong path.

If probing at two points in the wire, the meter's test probes must touch the metal inside of the wires in order to get a good volt reading. Touching the insulation of the wire is not enough. An alternative way of getting a secure reading is to connect to the metal terminals throughout the circuit such as the light bulb's terminals.

Tip About Wire Probes: As mentioned before, the test probes need to touch the metal part of the wires in order to get a reading. For the task of connecting to a wire, there are add on tools for your multi-meter to make this job easier. These add-on tools are called **wire piercing probes**.

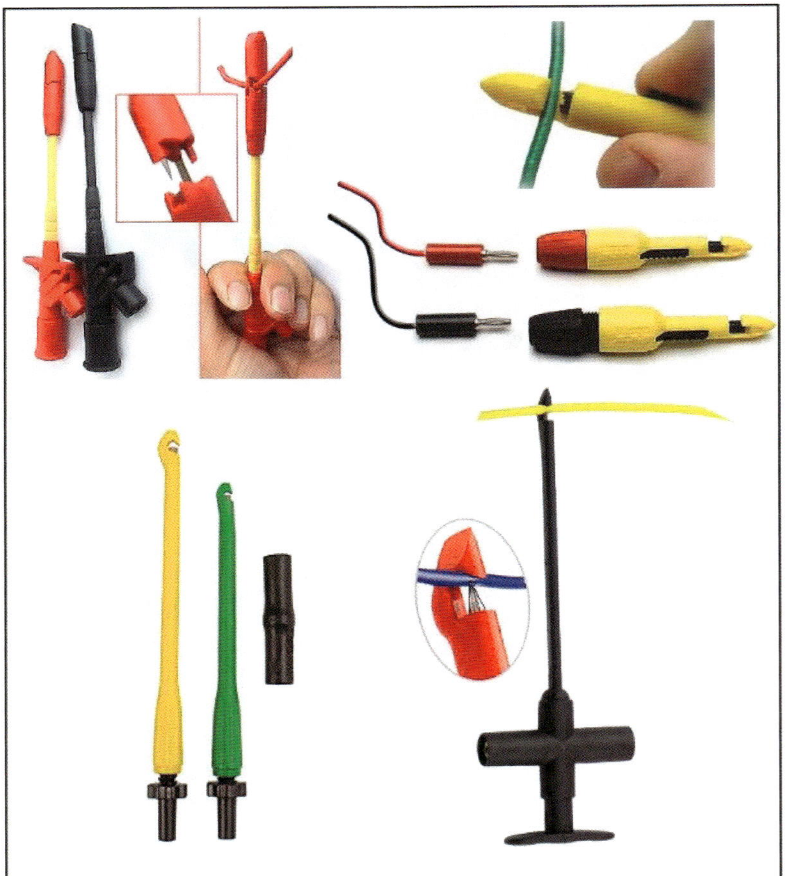

Different Kinds Of Wire Piercing Probes

(This illustration shows different examples of how to connect your voltmeter probes to a wire for a secure voltage measurement. You must make contact with the metal inside the wire to get a good voltage reading. The top image shows the needle tip of a test probe, piercing the wire to make a secure connection. The other images show other kinds of wire piercing probes that make this job easier)

One of the biggest advantages to these tools is the fact that they don't require you to hold the probe in place to secure a good contact and measurement. Just be sure that once you are done testing, you put a piece of electrician's tape over the tested area to cover the hole you made during testing.

If you are uneasy about piercing a wire for testing, another option is to use the metal terminals at the ends of the electrical parts in the circuit as test points for your meter's test probes. This is the better option to use when the entire circuit is easily accessible. I will elaborate more on this later in the following chapters..

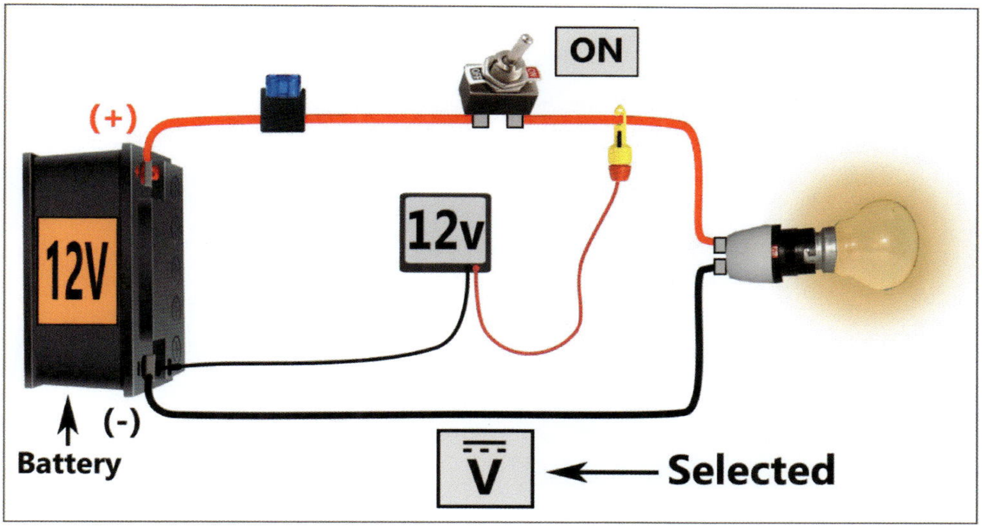

(Voltage testing using a piercing probe.This allows for a more hands free testing once the piercing probe has been installed. You can easily test various points in the circuit to determine if there is voltage where there should be.)

Battery Checks Using a Voltmeter:

One other basic thing you can use your voltmeter for is to check the voltage that an electrical power source such as a battery may have. A battery as well as other power sources usually have a voltage rating on it that you can compare your readings to. If after checking the battery you notice that the voltage is well below the rating that it should have, it is considered to have low charge. If the battery is not rechargeable, and the voltage is low then the battery needs to be replaced.

Note: Many batteries when low in voltage can be recharged with a charger to restore the proper voltage to the battery. You can also monitor this charging process by connecting a voltmeter to the battery while its charging.

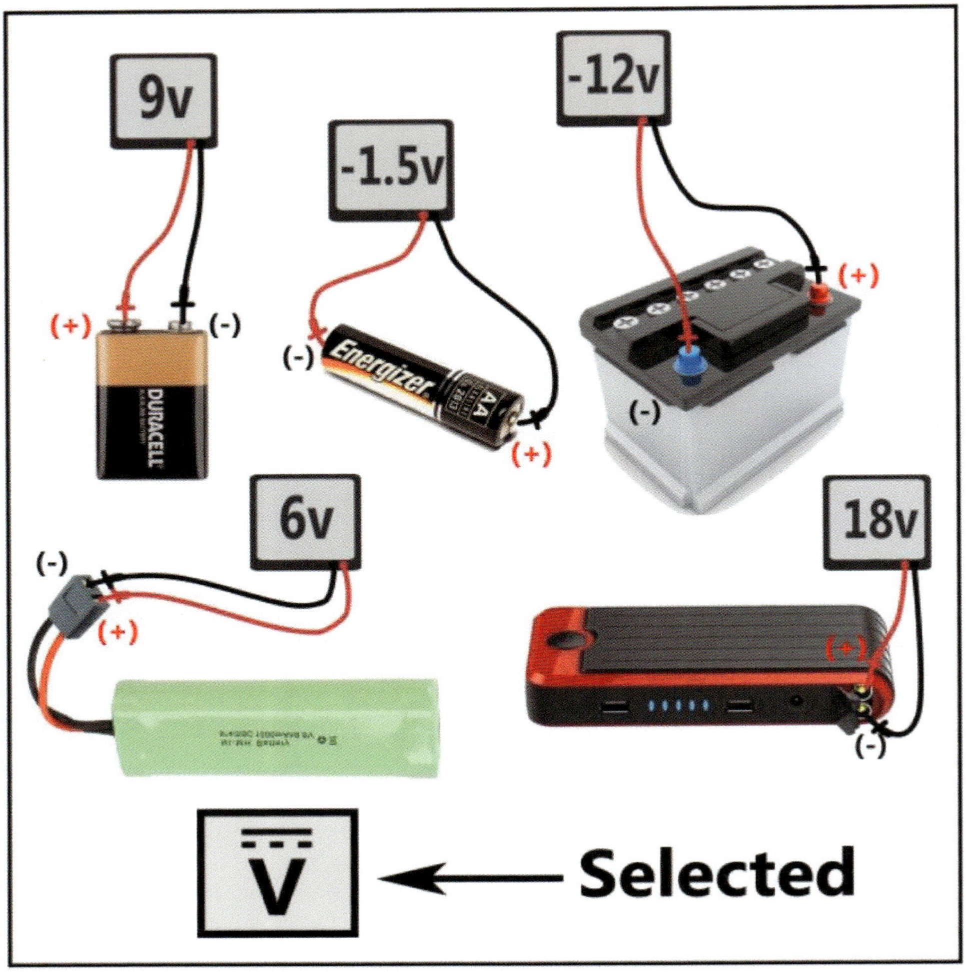

(Using the voltmeter to test various different types of battery for their available voltages. There are always two terminals to a power sources. The terminals are where you will place each test probe of your meter to check the voltage.)

In the previous image, notice that some of the voltmeter readings were negative voltages. Do not let that confuse you. This happens because the test probes are reversed. The polarity of the test probes was reversed by placing the red probe to the negative of the battery and the black probe to the positive of the battery. No damage will be done to your meter if you reversed the test probes during testing in the volt settings, you will simply get a negative sign on your voltage readings.

Ch.3: AC & DC Amperage

What Is Amperage?

Amperage is the unit of measure for the amount of electricity that is actually flowing throughout an electric circuit. It can be compared to the amount of water flowing throughout a water circuit. Do not confuse amperage with voltage. Although they are related and similar they are not the same thing. Remember that voltage is electrical pressure and that amperage is the actual amount of electricity flowing throughout the circuit because of the electrical pressure. Amperage cannot flow without voltage to push it throughout the circuit. An amperage measurement allows us to see the amount of electricity flowing inside a circuit. Amperage is also more commonly called **current**, **current flow**, **amps** and even **electrical flow**.

Considerations When Testing For Amperage:

Before we can perform a test for amperage or "amps", we must first take a look at how the **amp setting** differs from all the rest of the settings on the multimeter.

The first step to performing an amp measurement is set the meter to the appropriate amp setting and then place the red probe in the meter socket that matches the exact amp symbol that you have chosen. If the power source for the circuit is an AC type power source you will choose the AC amps setting on the meter. If the power source is a DC power source the meter should be set to read DC amps.

The second step is to turn OFF the circuit and then create a break somewhere in the circuit that you are testing. This can be done either at a connector or somewhere that is easily accessible to disconnect. Anywhere in the circuit will work but you should choose a place on the circuit that is easy to disconnect and then reconnect later.

The third step involves installing the ampmeter in between the break you created in the circuit. By doing this, you are actually making the ampmeter a part of the circuit. This allows electricity to flow through the ampmeter once the circuit is turned ON. The meter then measures exactly how much electricity is flowing through the circuit. For this reason, the ampmeter may sometimes be referred to as an electrical flow meter.

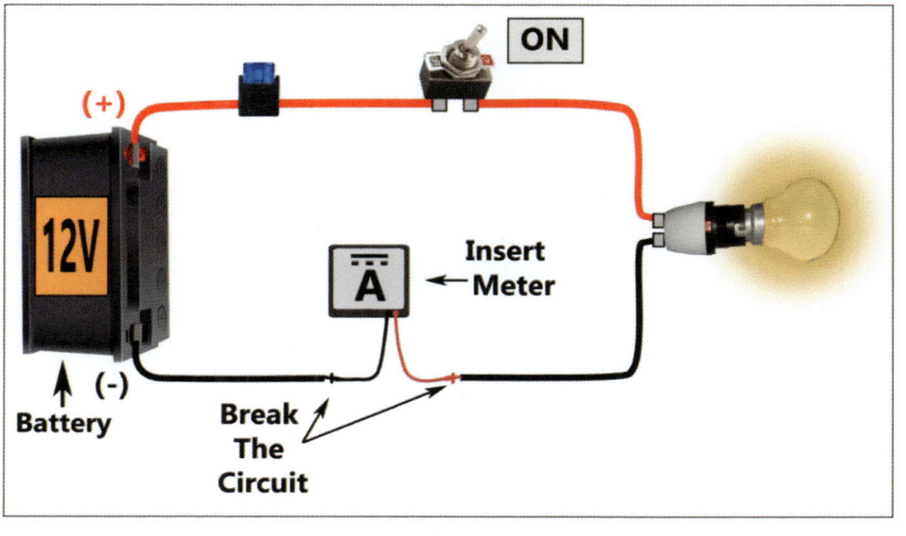

(This illustration shows how the ampmeter is installed for testing. There are three steps to performing a current test. First, you set your meter to the desired AC or DC amp setting and connect the test probes in the correct socket of the meter with the matching amp symbol. Next, you must turn OFF the electric switch and create a break in the circuit. Third, you install the ampmeter in between the break and turn the circuit back "ON" to check the current flow reading.)

With this setup you are completing the circuit through the meter. With the amp meter installed you can monitor how much electricity is flowing in a live circuit.

Note: When installing the amp meter into the break in the circuit, it does not matter what test probe goes to which side of the break. Either way will work and no damage will be done if you install the meter probes reversed.

Now lets take at the a look at some quick examples of the AC and DC amp settings in action.

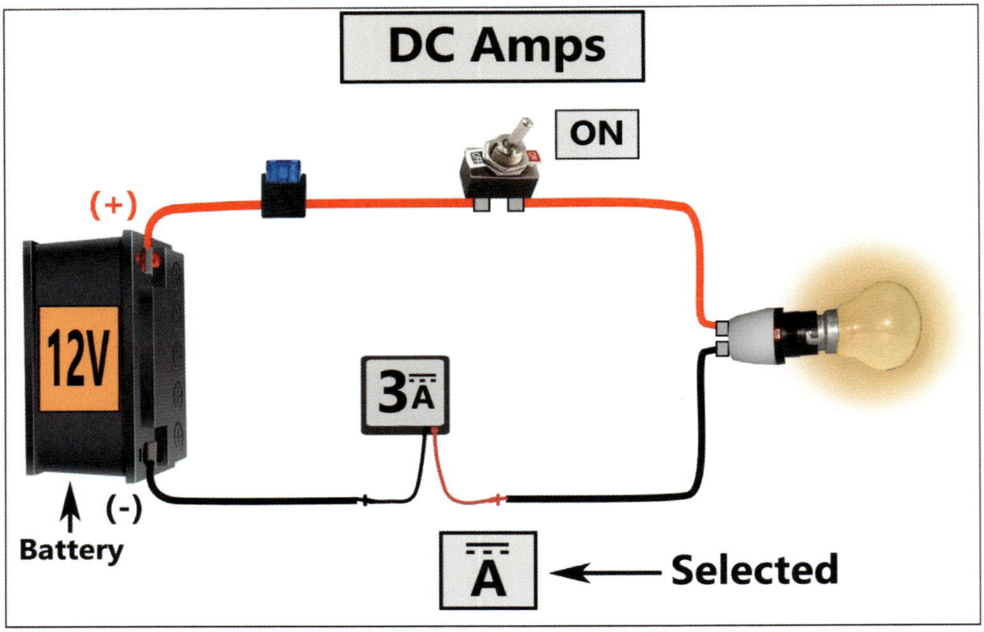

(**DC Amps:** In this picture, we see our original circuit with the ampmeter installed. The dial setting on the multimeter is set to DC amps and the amp meter has been placed in between the circuit as before. With the circuit ON, we can get a reading in the unit of **DC amps** of how much electricity is flowing inside this circuit. In this case we read 3 amps for this circuit.)

Now let us see an example of the AC ampmeter in action.

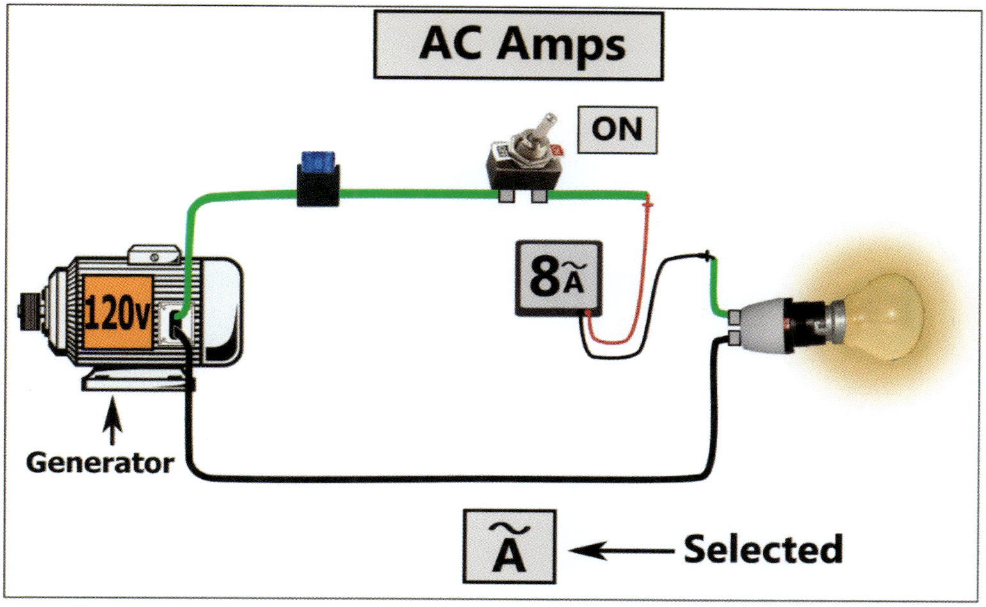

(**AC Amps:** In this picture, we see the AC version of the previous example circuit. The dial setting on the multimeter is set to AC amps now and the meter placed in a different location in between the circuit. With the circuit ON, we can get a reading in the unit **AC amps** of the amount of electricity flowing in the circuit. In this case, we read 8 amps on this circuit.)

As you have seen from the examples, testing for AC amps or DC amps using a multimeter is basically the same process. You break the circuit where most convenient, then you install the meter in between the break and then turn the circuit back ON to get your reading.

Note: Remember that it doesn't matter where the circuit is broken to install the amp meter because you will get the same reading anywhere in the path of the circuit. The same amount of amps will always be flowing in any part of the circuit.

Milliamps

For the sake of understanding the additional symbols that will be appearing on your meter, I have added an example of where an amp meter reads in milliamps instead of amps. Keep in mind that milliamps is not a new unit, it is still under the unit of amps. Milliamps is just a fraction of what an amp is in value. One milliamp is one thousandth of an amp, meaning that you need one thousand milliamps to make one amp.

Lets take a look at a current test on a circuit where the battery and load of the circuit are smaller than in the usual circuit we use for testing.

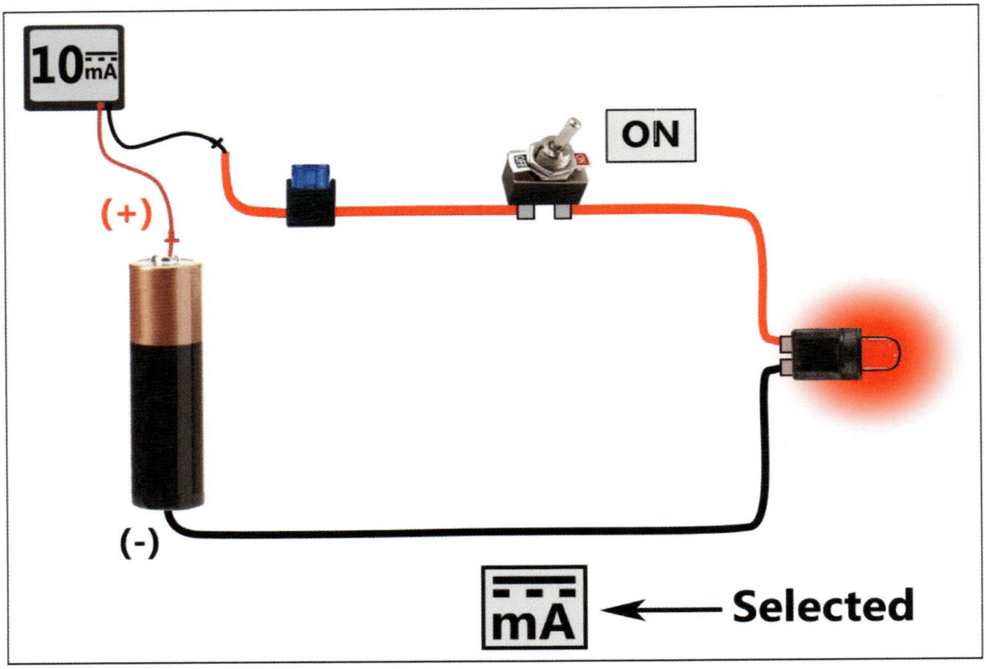

(In this circuit, we have our meter connected for reading amps just as usual. We see here that with a smaller battery and a small LED light we also get a small amperage reading. The meter reads 10 milliamps of electrical flow for this circuit.)

If you own a manual ranging multi-meter this is where you would select the mA setting on the dial. If you have an auto ranging multi-meter the meter will adjust the range automatically.

Note: In any of these settings you are dealing with measuring amperages or amps. Whether it is AC or DC amperage or ranging from over 1000 amps to a few milliamps, the hook up to a test area for measurement is always the same.

Breaking The Circuit

If only one circuit is connected to the power source, one of the easiest places to create a break is at the circuit's power source connections. Otherwise, if there is more than one circuit connected to the power source another option is at the connectors in the circuit which often are designed to be disconnected and reconnected with ease.

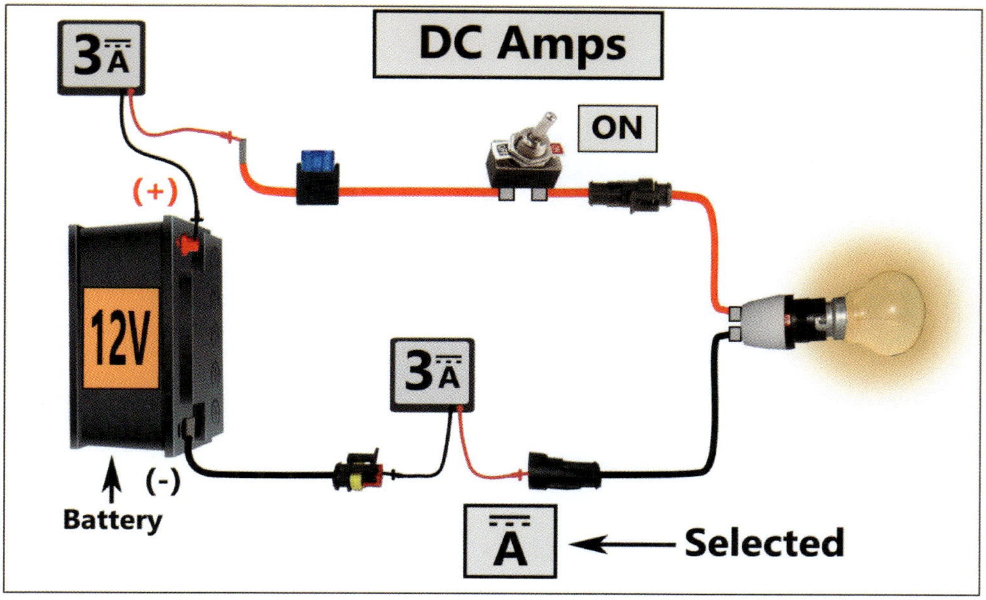

(Electrical circuits commonly have connectors or clamps that are designed to serve as a connection method for all the electrical parts of the circuit. These connectors and clamps can be disconnected and used for current testing, if easily accessible.)

The best location to choose for amp testing is always in the place where it is easiest to break the circuit to connect your meter to. Another easy place to connect your ammeter for testing is at terminals of the parts of the circuit. This is done by disconnecting only one side of the electrical part and then completing the circuit with your amp meter.

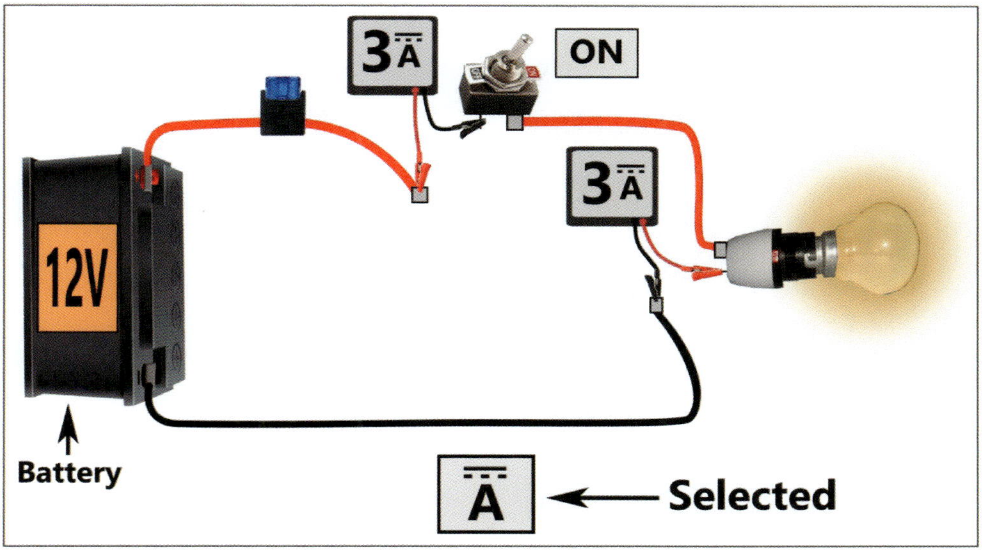

(An easy place to connect your ammeter is at terminals of parts of the circuit. This can be done by disconnecting one side of the electric part and connecting the meter in between the break. Notice the clamp style test probes being used for easier grip and setup.)

Precautions For Connecting To The Fuse Terminals:

One location I have commonly seen technicians connect their amp meter for testing is at the fuse's terminals. This is done by removing the fuse and connecting the meter in its place. The reason this is done is because of the ease of being able to test a lot of circuits for amperage fairly quickly at the fuse panel by removing the fuses and installing your meter.

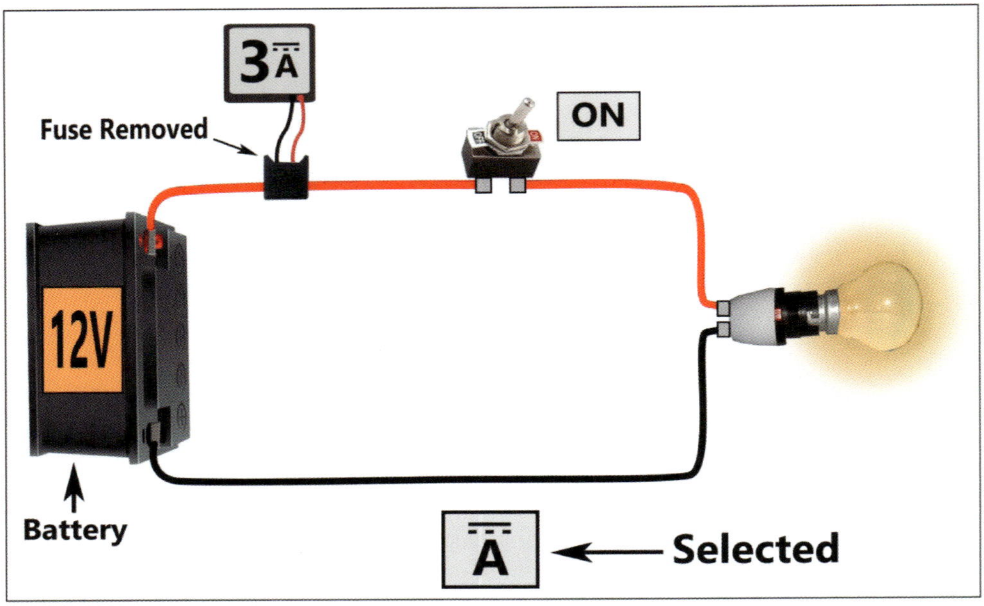

(Amp testing using the fuse terminals. The fuse must be removed and the meter installed in its place.)

The main risk in using the fuse terminals for current testing is to understand that the amp meter will now be in place of the fuse. This means that there is no proper circuit protection against a short or a power surge. If any one of these conditions where to happen while testing you would likely damage your meter or blow the internal fuse in it. A worst case scenario is permanent damage to a circuit. Be cautious when using the amp meter. If you are using this location for hookup of your meter be mindful not to accidentally cause a short while your amp meter is in place of the fuse.

Cautions When Using the Amp Setting:

The main problem to worry about when using your ampmeter is the possibility of causing an accidental short by hooking the amp meter up incorrectly. Usually, if you were to create an electric short by hooking the amp meter up wrong, this will blow a circuit's fuse or the multimeter's internal fuse.

Always remember than while current testing, the ampmeter is actually completing the circuit. You can think of the ampmeter as if it were two jumper wires with a measuring gauge in between it. This can help to understand how you can actually hook the amp meter up wrong.

For example, you should **NEVER** place an ampmeter across an electrical device in a live circuit or across any power sources either. If you do, YOU WILL CAUSE A SHORT. Worst of all, you might destroy your multimeter if you create a short this way. Be careful.

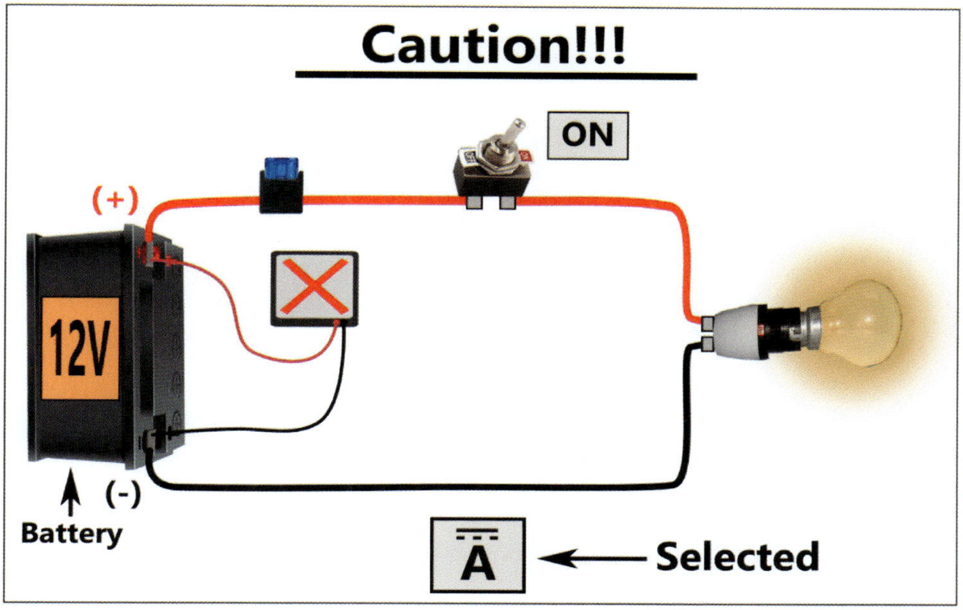

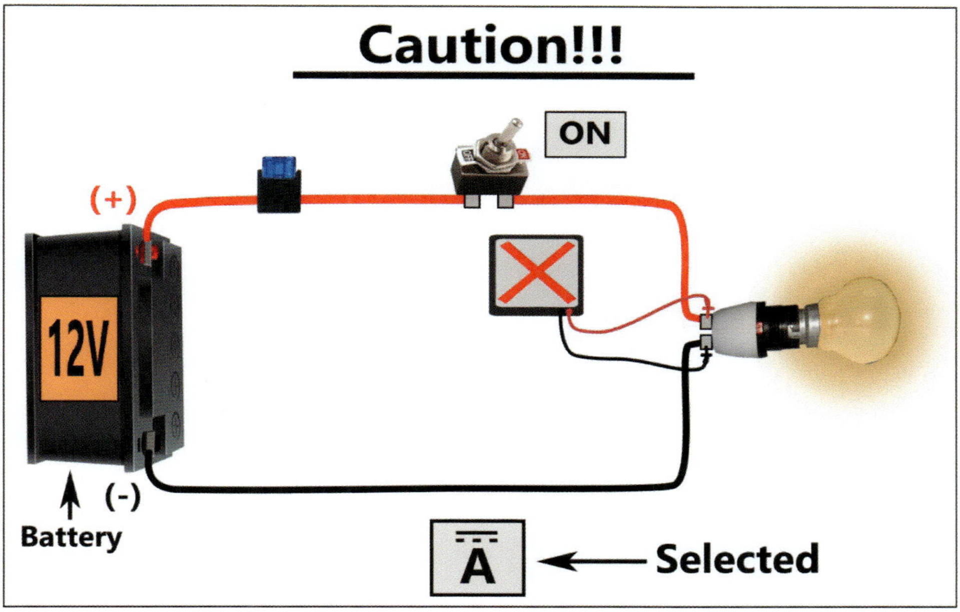

(Two of the most common beginner mistakes involves using the ampmeter like if it were a voltmeter. Creating a short across the battery and/or creating a short across the electrical load of the circuit. Remember that an ampmeter is like a set of jumper wires with a measuring gauge on it.)

Basically what you are doing in either of these example is creating a short directly from the power side of the circuit to the negative side. If you "jump" the battery or the electrical load using your amp meter you would be creating a direct electrical short from positive to the negative. If you are lucky the short might only blow a fuse in your meter. If you are unfortunate to be working on a circuit that does not have a fuse and your meter also doesn't have a fuse, you may accidentally cause permanent damage to the circuit or destroy your meter. Do not bypass a power source or an electrical load. Please be careful not to make this mistake.

If you are interested in more about electric shorts, how they affect an electric circuit and how to find and fix these kinds of issues please check out my other book "How To Find Electrical Shorts"

Uses for Current Testing:

One of the reasons why we would check for amperage of a circuit is because it allows us to monitor how well the circuit is performing when turned ON. If the amperage is too low or too high compared to the proper amperage value for the circuit this can be a sign that a problem may exist with either the electrical device or something else in the circuit. If we already know the actual amperage that the circuit should have, we can use the measured readings to compare to the known good values to confirm if there is a problem in the circuit.

The basic amp meter is mainly used in prototyping labs and manufacturing facilities to confirm that a device is performing as it was made to perform once installed in a circuit. The ammeter may also be used during the construction of a circuit to confirm the actual current draw so that a properly rated fuse or a circuit breaker is installed to protect the circuit in the case of electrical surges.

Determining Which Circuit Is ON or OFF.

Another job that can be done with the amp meter is to check which circuit is on when you are in a place where you can't visually see if the circuits that are ON. In many industries, the circuits you are working on can be very large in length and you wont be able to see all the parts on the entire circuit.

Just as before, if you connect the amp meter to a circuit and get an amperage reading this means that the circuit is ON and current is flowing. If there is no amperage reading this means the circuit is OFF or there is no current flowing.

Confirming Open Circuits:

One other basic use of the ammeter for troubleshooting purposes is to confirm open circuit problems. If there was an open or break in the circuit, we could confirm that this is the issue by using current testing. If we were to read 0 amps with the electrical switch ON, this means there is no electrical flow in the circuit. This can either be caused by a bad switch, a bad device or another kind of a break somewhere in the circuit.

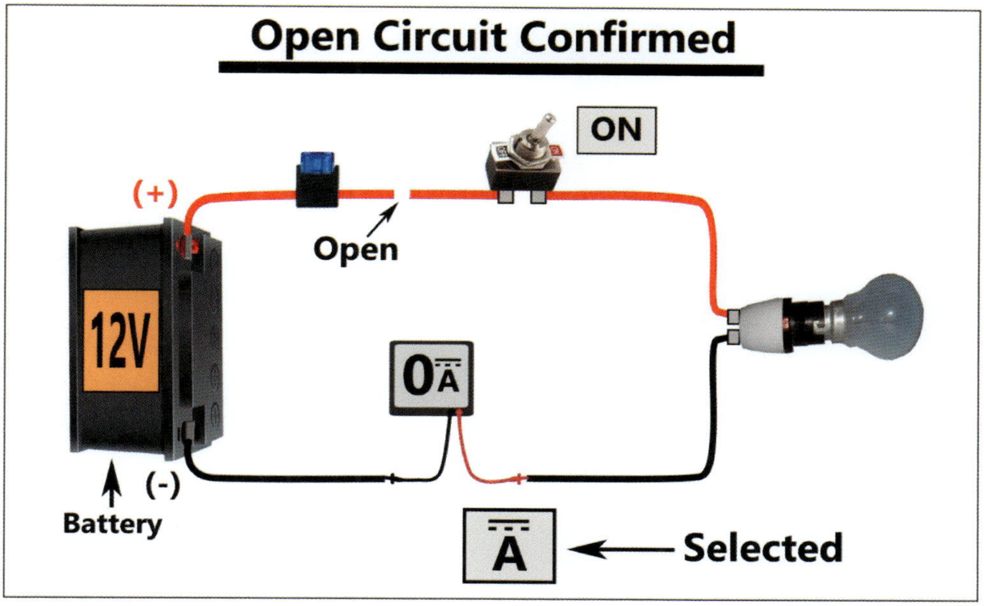

(In this circuit, we have a light bulb that wont turn on. By connecting our amp meter to the circuit we can confirm the type of problem that there is. The 0 amps reading on the meter with the electric switch turned ON confirms that there is an open somewhere in this circuit. After confirming the problem, further testing would be necessary using the voltmeter to track down the exact location of the open.)

As you may have already noticed, the amp meter setting as a tool for troubleshooting is limited to being able to only confirm whether the correct current is flowing when the circuit is switched to the ON position. This is the reason why the amp setting on a regular multimeter is usually reserved for basic checks and not as much for finding a problem.

Ch.4: Resistance, Continuity, Diode and Capacitance Function:

Ω	**-RESISTANCE, OHMS**
•)))	**-CONTINUITY TESTER**
	**-DIODE TESTER**
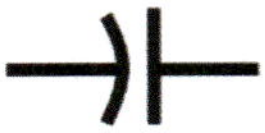	**-CAPACITANCE TESTER**

The following terms are important to remember before using any of these test settings.

Resistance:

Resistance in electrical terms refers to an electrical part's resistance to electricity when the part is installed in a circuit. Anything that can disrupt, slow down or prevents the flow of electricity is considered **electrical resistance**.

Everything electrical has some degree of electrical resistance inside of it, regardless of whether it is an electric switch, a lamp, a fuel pump, a heater, a connector or even the electric wiring in the circuit. Basic parts such as wires, connectors, switches, terminals and fuses also have a resistance to electricity, although it is a very tiny resistance and usually ignorable when compared to other electrical parts.

When electrical parts are designed by engineers they are built to have a specific amount of resistance. When circuit designers make a circuit they take these resistances into account to create a properly working circuit. The main resistance of an electric circuit is generally the electrical load(s) of the circuit.

It is the job of the resistance meter (or ohmmeter) to measure the resistance inside an electrical part to check for proper resistance. If too much resistance exists inside the electrical part when compared to what a new one should have, then it is considered bad. When the basic parts used to build a circuit are "Good" they usually have a very tiny resistance that doesn't interfere at all with the electrical load of the circuit working properly. But when these basic parts become bad, they can have too much internal resistance and cause performance problems to the circuit.

Quick Note: The symbol for resistance next to your reading on the ohmmeter is called the **ohm**.

How The Ohmmeter Measures Electrical Resistance:

It is important to understand at least a basic idea of how each of the meter settings inside the multi-meter works. The **ohmmeter** measures the resistance of something electrical by actually applying a small constant voltage (usually 0.5 volts) to the test piece and then measuring the resulting current created by this applied voltage. By sampling the current and knowing the applied voltage, the meter can calculate the resistance using these two already known values. The meter internally calculates the resistance of the piece by using the value of the voltage applied and the current created and then placing these values into an electrical formula known as **OHM's Law**.

Ohms Law In A Nutshell

Ohm's Law is one of the first basic laws of electricity that changed our understanding of how electricity works. The law states that voltage, current and resistance are all related.

The relationship between voltage and current is somewhat identical in terms that if voltage goes up in value, the current goes up in a similar amount as well. The relationship between voltage & current vs. resistance is inversed. This means that when resistance goes down, voltage and current go up. If resistance goes up, voltage and current will go down. It is now known that the resistance of a part is what will ultimately determine how much electricity can flow when installed in a circuit.

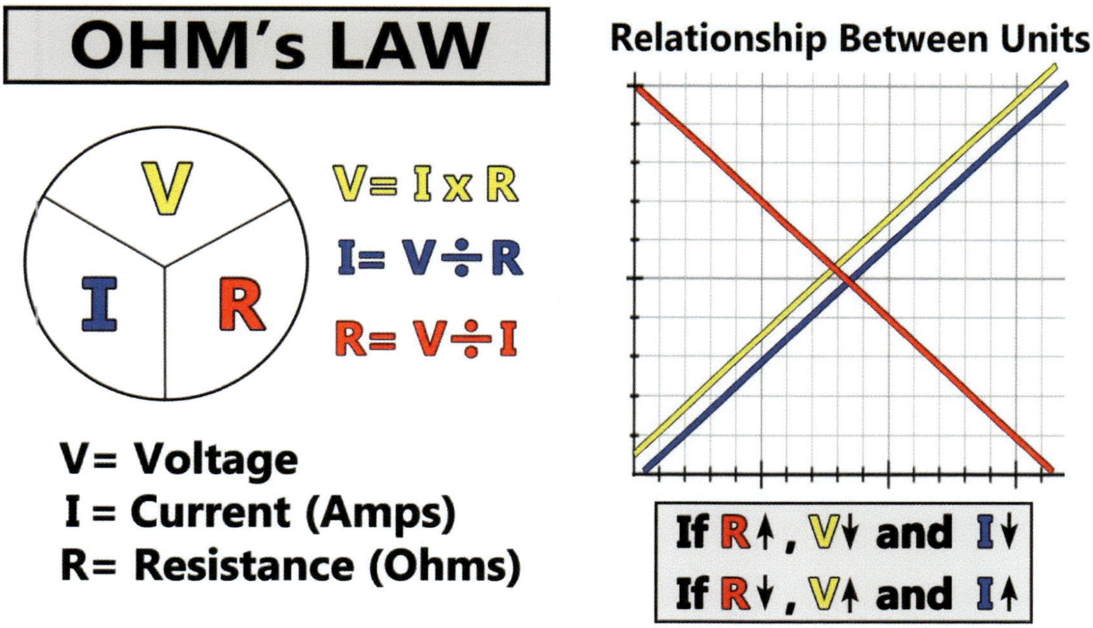

(This image shows the formulas in Ohm's Law. By knowing two of the values you can calculate the third value using one of the formulas. The line chart represents the relationship between voltage, amperage and resistance.)

The key takeaway from this law is that the values of voltage, current and resistance are all related and will affect each other depending on each others changes. By understanding this relationship, we can understand how the ohmmeter uses OHM's Law to calculate its measurement and how a circuit can be affected when the resistance of one of the electrical parts goes up.

The current flowing into the ohmmeter during testing will change depending on the amount of resistance of different test pieces. The applied voltage to the test pieces remains constant at 0.5 volts. This allows the meter to calculate the exact resistance depending on the current flow value that the test piece allows to flow when the specific voltage applied.

How to Use an Ohmmeter:

In order to start measuring the resistance of a test piece, it must first be disconnected and then tested separate from the circuit it was in. The resistance test is more of a bench test for electrical parts and it isn't used for diagnosis while the circuit is live and working.

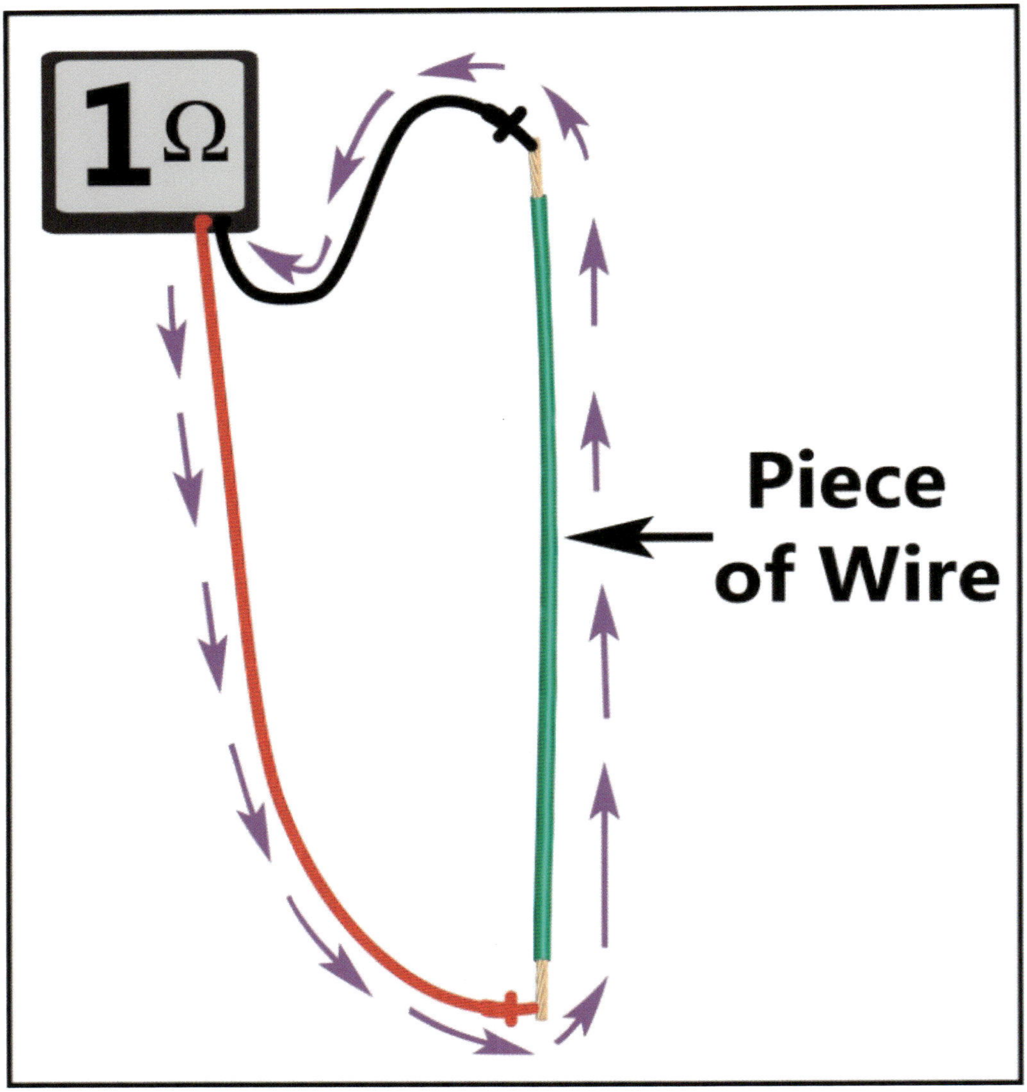

(**Resistance Testing Example #1:** Resistance testing a section of wire. Remember to test only with circuits off or the electrical piece disconnected. In this example, I have shown the electricity that the ohmmeter applies during testing (arrows in purple) only for illustration and teaching purposes.)

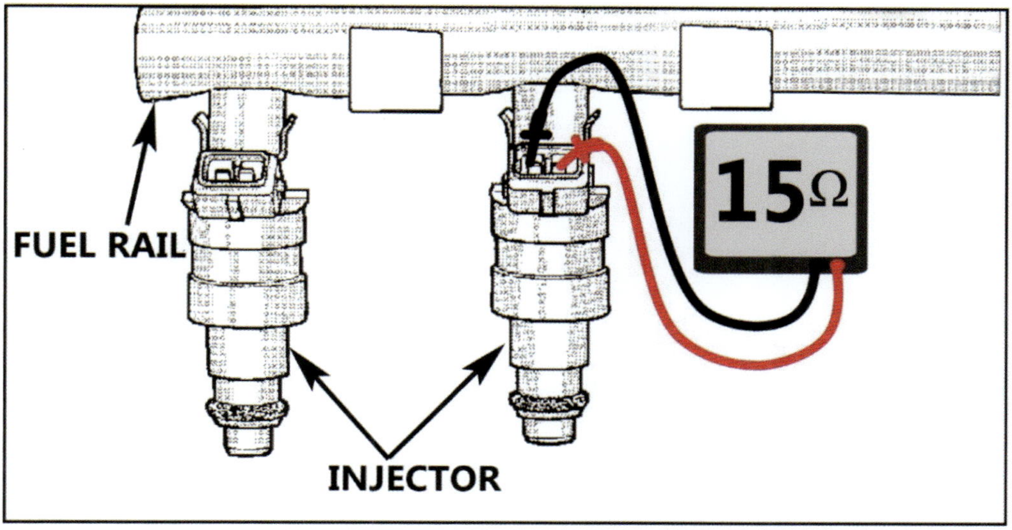

(**Resistance Testing Example #2:** Resistance testing an electrical solenoid valve that is being used as a fuel injector. Remember to test only with the circuit off or the electrical piece disconnected from the circuit.)

(**Resistance Testing Example #3:** Checking the internal resistance of a light bulb)

(**Resistance Testing Example #4:** Checking resistance of an audio speaker.)

Testing your Ohmmeter for Internal Damage:

One thing to remember when using the ohm setting is to NEVER use the ohmmeter on a live circuit. Always disconnect the part to be tested first. If you accidentally do test the circuit live, you will blow the meter fuse or even damage your meter. Be careful not to do this. If you do not know whether your ohmmeter is actually working or not you can test it by touching the two test probes together while set to the ohm setting.

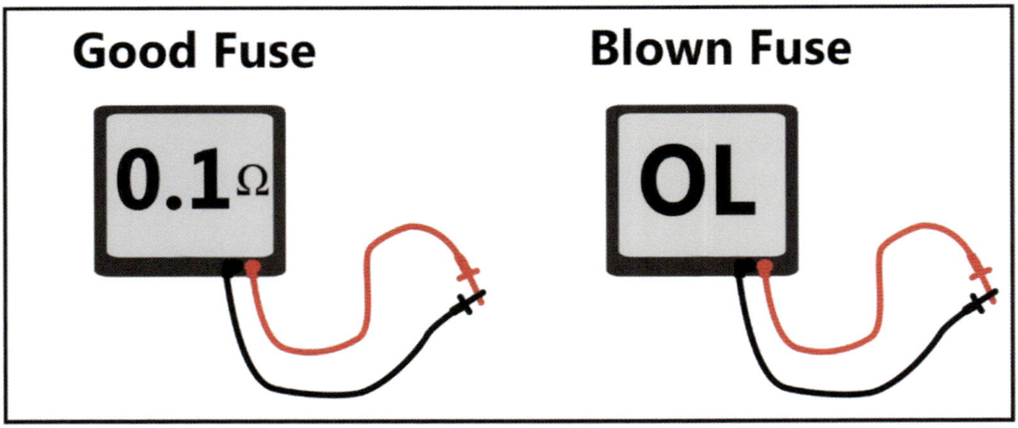

(This is how you would know if you accidentally blew your fuse on your meter or damaged it with improper use. Place the meter in the ohm setting and touch the two probes together to test your meter internally. When touching the two test probes together the meter should have a small ohm reading if everything is still working okay. If something was accidentally damaged, you would get an overlimit "OL" reading or a blank reading.If your meter is equipped with a fuse, open the meter and replace the fuse to restore operation.)

Testing For Bad Parts Using Resistance Testing:

The OL reading on the ohmmeter means "open" or in other words the piece being tested is internally disconnected or broken. There is no complete path through the part being tested. This is useful to know when testing parts because an OL reading would mean that the part is internally broken. The electricity that the ohmmeter applies to test a part never returns through the other probe because the path of electrical flow is internally broken within the part.

In an electric switch there is an additional step to consider when testing for resistance. The completed path inside the switch is intentionally broken when it is in the OFF postion. An ohmmeter test for a good switch should result in an OL reading in the OFF position and a very low resistance in the ON position.

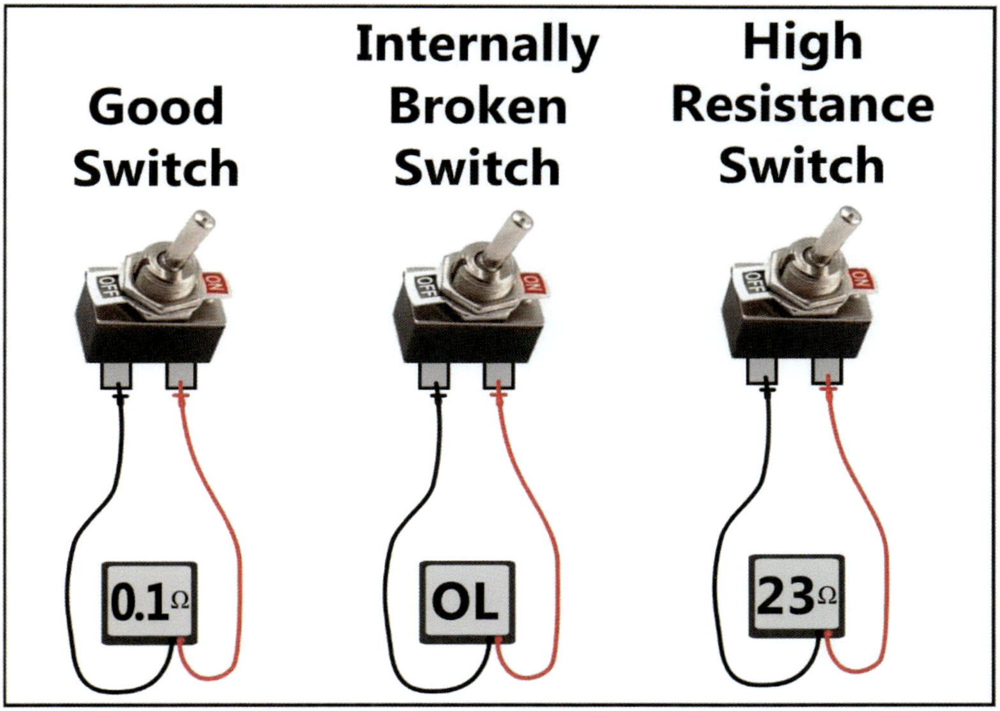

(Examples of three possible results from testing an electric switch for resistance when it is in the ON position. The first image shows a low resistance reading from the switch indicating a good switch. The second image shows an OL reading from a switch that means it is internally broken. The final reading shows a switch that is worn out and has too much resistance inside it.)

Any part with too much resistance can cause performance problems to the circuit that it is installed in. The second and third switch readings in the previous example indicate a bad switch that needs replacement.

Testing For Shorted Parts Using Resistance Checks:

Another way that an electrical part can fail is by becoming internally shorted. When this happens, the shorted electrical part can cause all sorts of unique problems to the circuit that it is installed in. This condition can cause a blown fuse, strange behaviors of the circuit, permanent damage to the circuit and even start a fire. There is a lot more to explain about electrical shorts and how they affect a circuit and for this reason I have dedicated an entirely separate book to the subject. But for the sake of understanding the ohmmeter let us explore how to test for a shorted electrical load.

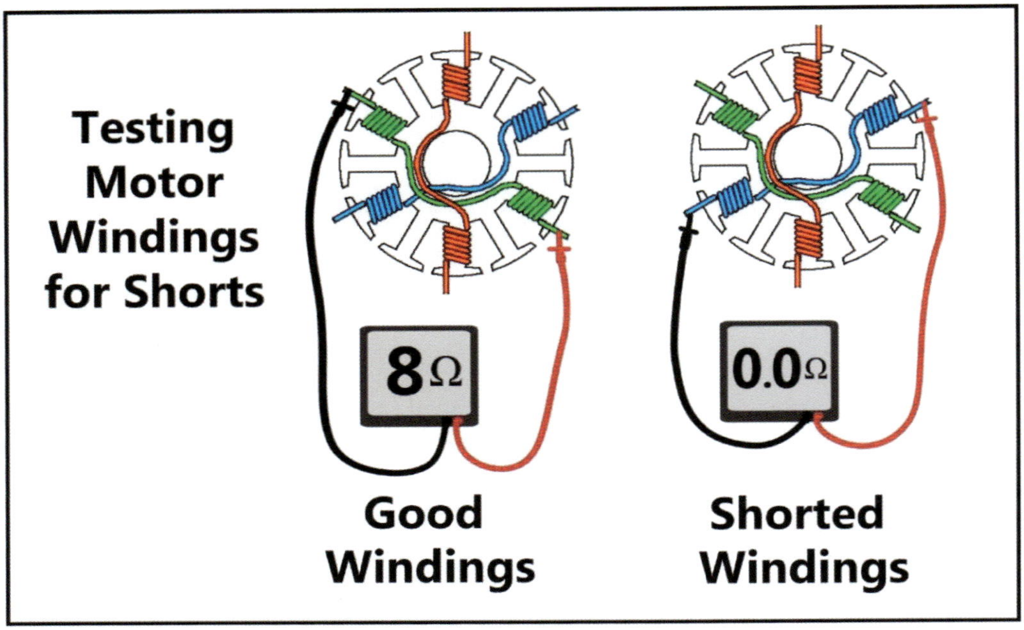

(Here is an example of how you can use the ohmmeter to test the internal part of a motor for an electrical short.)

You can see that the green coil windings inside the motor measured a normal reading of about 8 ohms. Since the blue and red coil are identical to the green coil they should all have the same resistance. In the second example we notice that the blue coil winding had 0 ohms on the ohmmeter. This means that there is a low or even no resistance at all within the part being tested. This can only happen if the coil winding's rubber insulation has worn down and the coils are touching together and shorting out.

Now lets look at another example of a short in an electrical device.

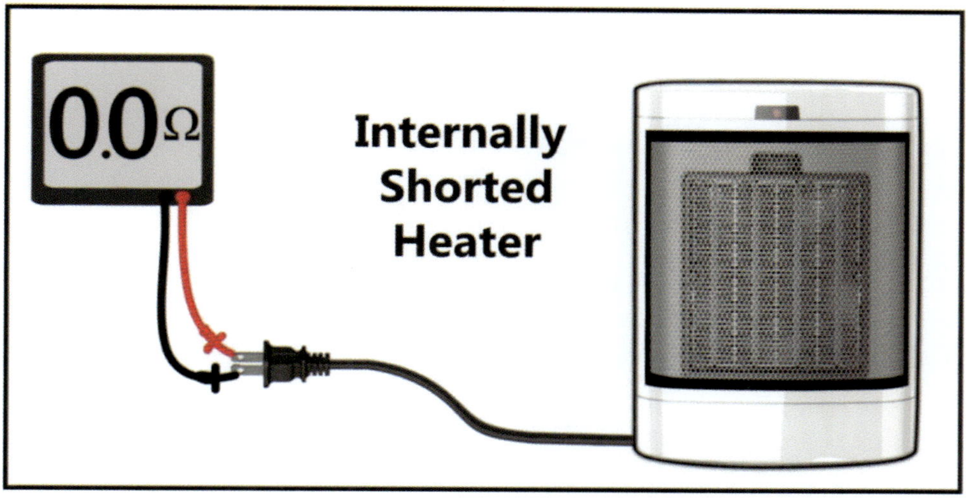

(Example of a heater being tested for resistance. The result was a 0 ohm resistance indicating that this heater is internally shorted.)

These are but a few examples of a part with an internal electrical short. A near 0 ohms reading on an electrical load would definetly indicated an internal short. Only an electric switch, a fuse, electrical clamps, wiring or electrical connectors should have little to near zero resistance when ohm tested. Every other electrical part should have the resistance amount stated in their electrical part spec sheet. If the amount of resistance is lower than normal when compared to what it should be, then the part being tested is shorted or partially shorted and must be replaced.

Testing Wires Using The Ohmmeter:

With the help of the ohmmeter It is also possible to test a wire and determine whether it is internally broken by using the resistance test.

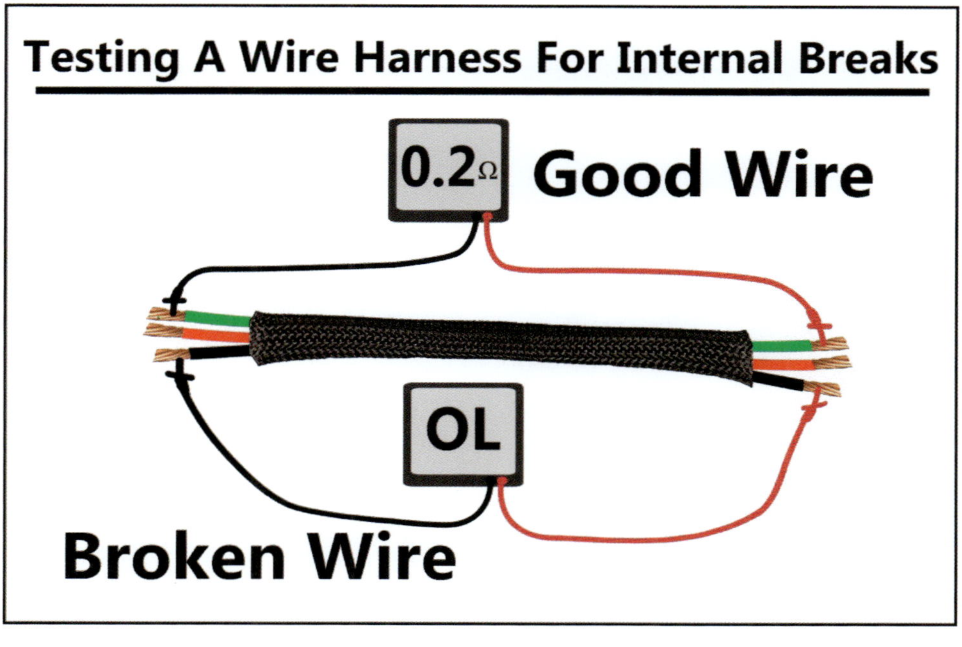

(This illustration shows how to test wires in a bundle for resistance. The same rules apply about needing to touch the metal part of the wire to get a proper reading. With this positioning of the test probes you are testing for resistance between where you placed the two probes. If the wire is good between the two test points, it will show a very low resistance reading on the meter. If the wire is internally broken between the two test points, the meter will display "OL".)

Important Note: The wire test is only accurate if the electrical wire is either completely broken internally or if the wire is 100% good without any missing strands. This wire test is not accurate if the wire is partially broken or has missing strands. The ohm test may show that a wire is good when in fact there are internally broken wire strands that could be the cause of why a circuit isnt working properly.

Resistance testing will not be able to catch an issue where the circuit isnt working because of some broken wire strands within the wire and for this reason the ohm test is limited in this area. I explain why this is and also teach a better method of testing wiring problems in my other book "How To Test Like a Pro."

Just as with any other kind of electrical measurements you take with a multimeter, a resistance measurement should be compared to a known good value. This can be done by comparing to the manufacturer values on the spec sheet, the resistance rating marked on the body of the device or by comparing to a part that is identical to the part that is being tested.

Keep in mind that all the resistance readings I have shown involving electrical devices being tested are not values that should be taken as a definite guide. The resistance readings can and will vary greatly from device to device and should be compared only to the exact resistance that the specific device is rated to have.

Electrical Resistors:

An explanation of the unit of resistance would not be complete without introducing the electrical resistor. A resistor refers to a device that is specifically designed to resist the electricity in the circuit that it is installed in. This resistance is added into a circuit on purpose to allow for a way to limit the amount of current that can flow within the circuit.

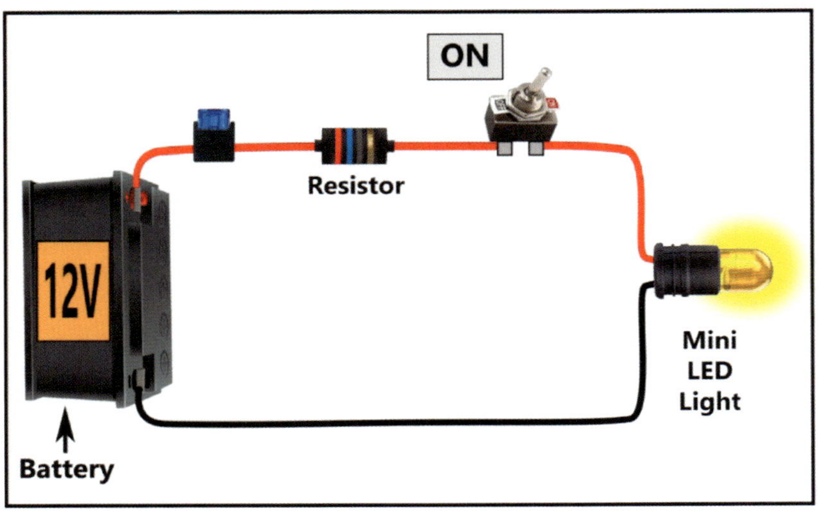

(Example of a basic circuit with a resistor installed inline to control the maximum allowable current that may flow.)

The electrical load in a resistored circuit will work along side the resistor, with the limited electricity that the resistor allows to flow throughout the circuit. The resistor is only used when there is a need to set a limit to how much current can flow and also protect the electrical load that may be unable to handle the full electrical potential that the power source can provide. The resistance inside of the resistor ultimately will determine the maximum amount of current that is allowed to flow within a circuit.

Note: A resistor can be considered a second electrical load on the circuit. Its main purpose being to reduce the current down to a level that will work best for the primary electrical load of the circuit. The primary electrical load may not have enough resistance by itself to limit the current in the circuit to a level that is acceptable. For this reason, a resistor is added.

Now lets take a look at an example of resistors being ohm tested to determine its internal resistance.

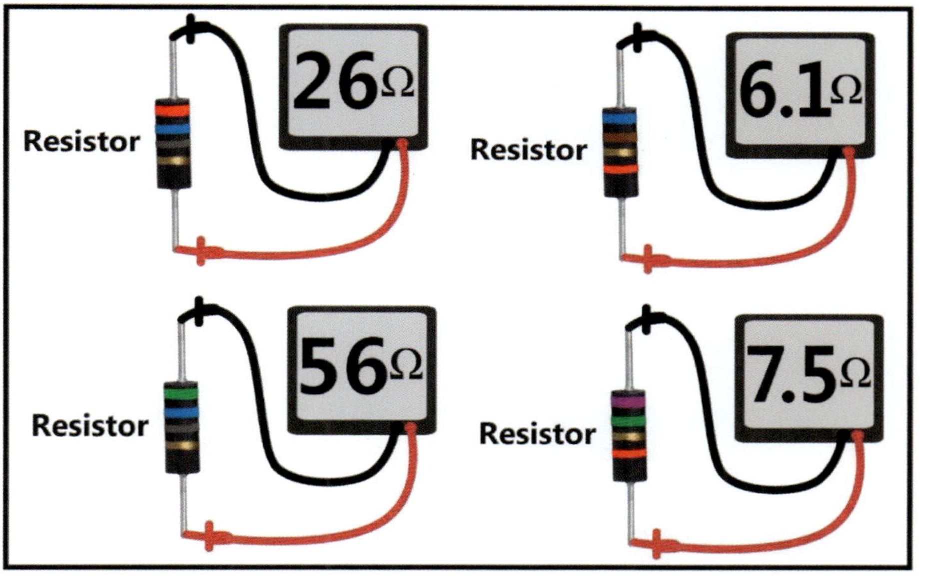

(in this example the ohm meter is testing a set of resistors to determine theirs internal ohm value. These types of resistors are coded with different colored bands on the body of the resistor that can be translated into a certain amount of resistance value.)

As a recap for the lesson on resistance, the basic parts that make up a circuit that don't involve resistors or other electrical devices should normally have a very low resistance. The electrical load(s) of the circuit including resistors, if used, should be the main resistances in the circuit in order for the circuit to work correctly. If there is ever a electrical part found to have a higher resistance that it was designed to have then this part should be considered bad and must be replaced.

The Diode Check Function:

A diode is an electrical device that only allows electricity to flow in one direction. It is like an electrical one way check valve that allows electricity to flow one direction but blocks electricity from going in the opposite direction.

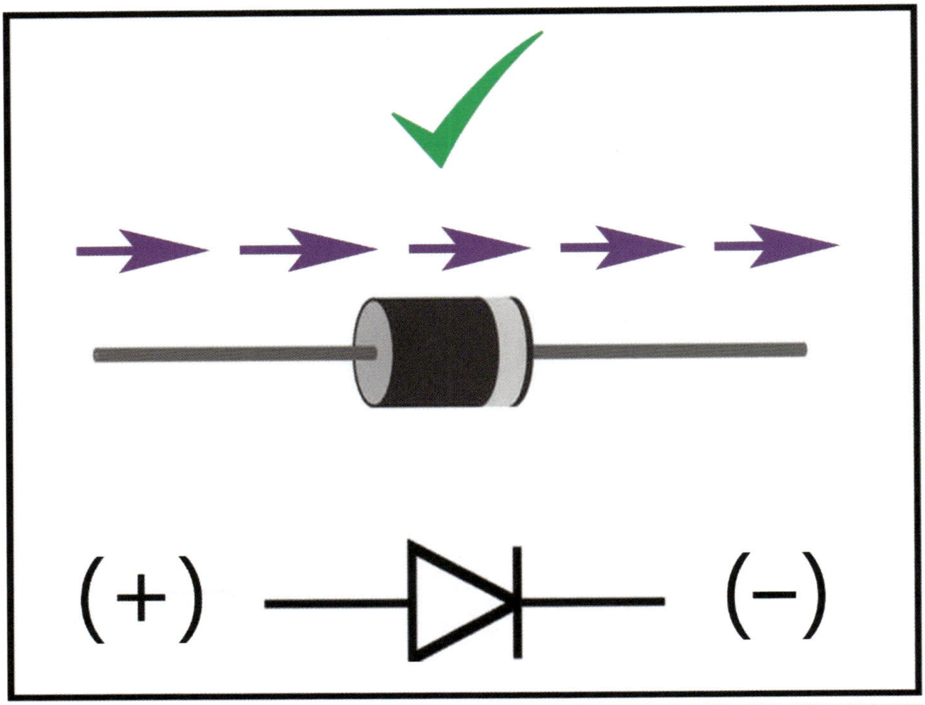

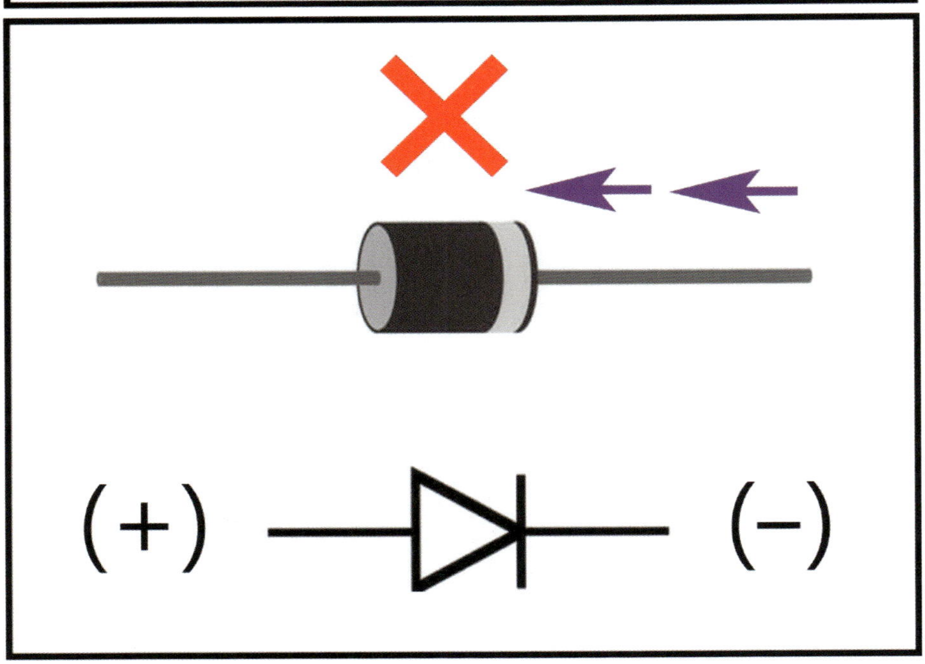

(A diode is designed to allow electricity to pass one way, yet block it from flowing the opposite way.)

Diodes are usually only used in AC circuits where electricity moves back and forth. AC electricity alternates flowing in both directions inside of the electrical circuit. This is why AC is called alternating current. DC electricity on the other hand only travels in one direction all the time.

A diode can make AC electricity flow in one direction just like DC electricity. The diode when bad will no longer be able to control the electricity to move in only one way. A bad diode allows electricity to go in both directions. This will cause many performance problems to a circuit if it wasn't designed to work on AC electricity. To test a **diode,** use the **diode check** function on your multimeter. The symbol for a diode looks like a triangle with a vertical line.

How to Perform the Diode Test:

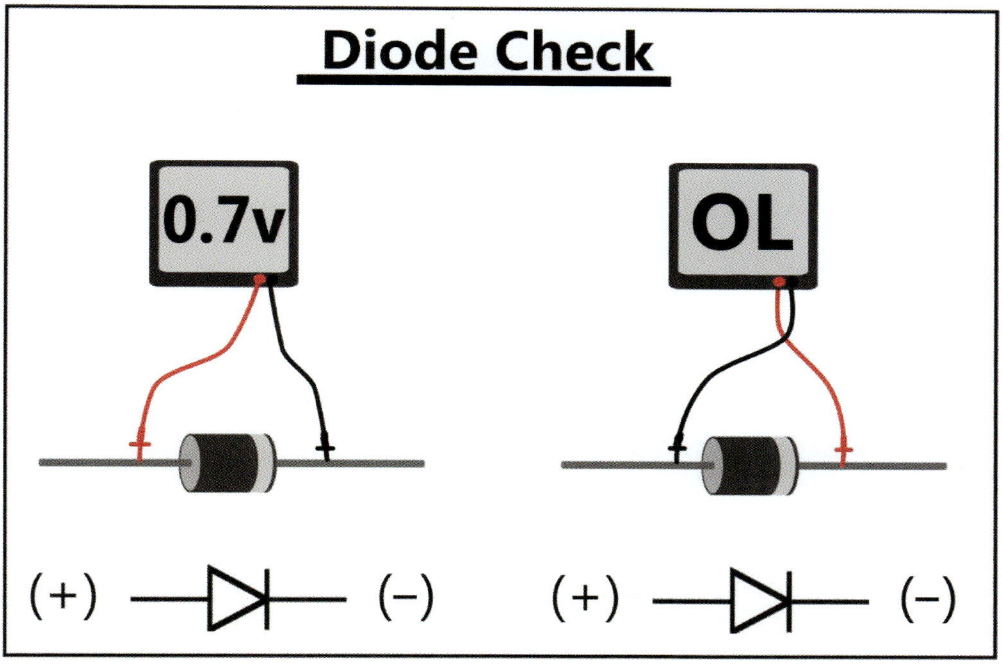

(In order to perform a diode check, you must first put the meter into the diode check function. The diode check is a two-part test. Start by probing each side of the diode to get the first reading, then continue on by reversing the positions of the test leads and retesting for the second reading. If the diode is good, one test reading should read near 0.5- 0.7 volts and the other test reading should read "OL" or overlimit.)

If you read a voltage both ways this indicates a **shorted** diode. If you read "OL" both ways this indicates an **internally broken** diode. Depending on your readings replace the diode if necessary.

The Continuity Check:

The word **continuity** refers to the completeness of a circuit or of an electrical part being tested. The continuity test is basically an ohmmeter with a buzzer. It is used as a quick check of the internals of something electrical to make sure they are not broken. When there is continuity inside of a part or between the two points where you are testing, you will hear a continuous beep, indicating completeness or continuity internally.

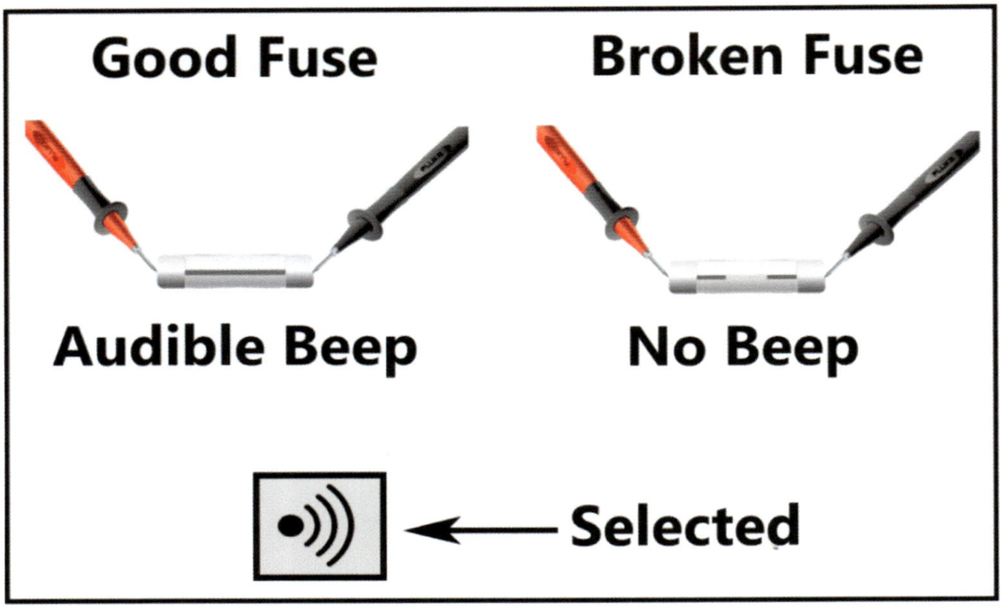

(The continuity tester is used in the exact same way as the ohmmeter. The only difference is that there will be an audible beep when there is continuity in the part being tested. You will NOT hear a beep if the part being tested is internally broken. In this image we have a see through cartridge style fuse being tested for continuity. The beeping is only heard on the fuse with the internals in good condition.)

This test can be applied to just about any other electrical part, including wiring. This test is only usually used as a quick check to see if something is completely broken internally.

In this case, the fuse could have been easily checked just by looking at it and seeing that it is was internally broken but other electrical devices are not always see through and this is one of the tests to use to find a problem.

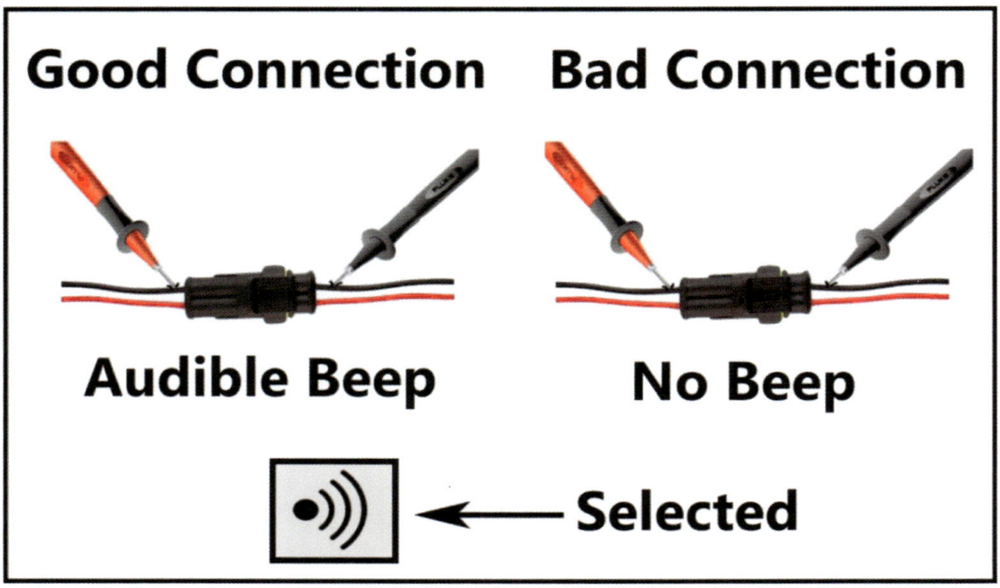

(This is an example of an electrical connector being tested with the continuity tester. The connector is not see through like the fuse in the previous example therefore testing is required. The connector is being tested for continuity by piercing into two points in the wiring.)

As you can see in the example, there are times where there is no difference in appearance between a broken part and a good part. This is why testing is absolutely necessary. In this image, the connector terminals that connects the two black wires together are included with the positioning of the test probes. The other terminals of the connector that connects the red wires would be tested in the same way. You can also use this continuity test when quick testing multiple wire connectors.

Reminder: Every ohmmeter test including the Continuity test is done with the part removed OR with the electrical circuit turned OFF first before testing.

Capacitance:

Capacitance refers to the storage capacity of a capacitor. A capacitor is an electrical device that is able to store electricity inside it, to then later on be used, somewhat similar to a battery. The capacitance test measures the capacity of electrical charge that a capacitor can store.

Before we begin tests on a capacitor let's take a look at what they commonly might look like..

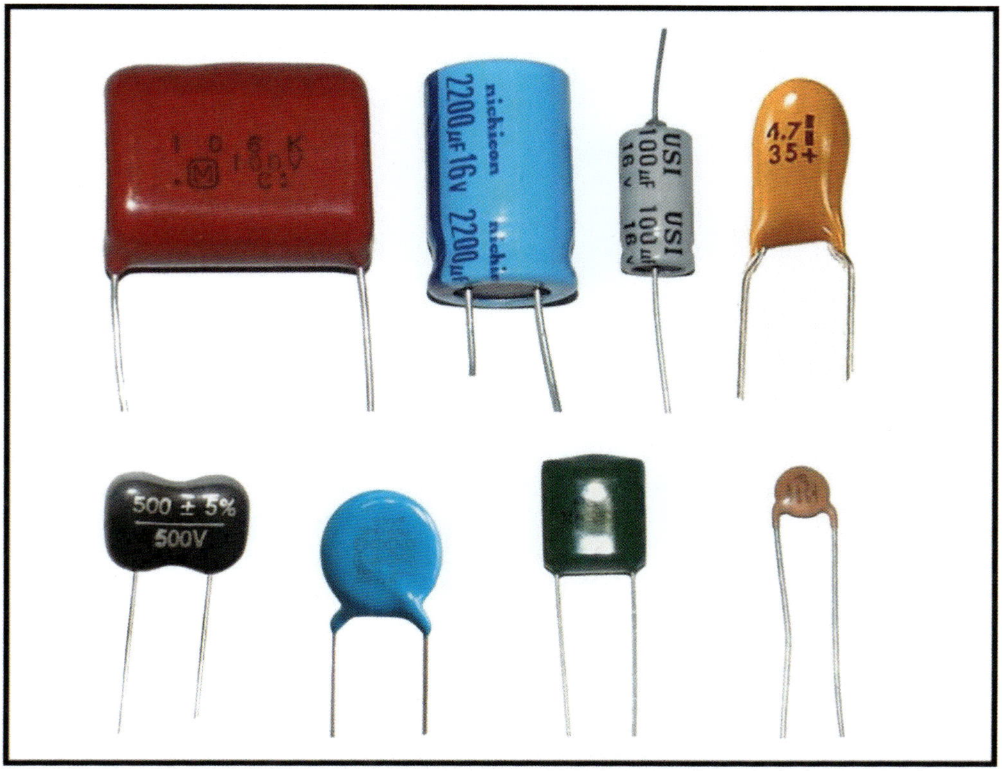

(This illustration shows different variations of what some **capacitors** may look like. They are made of many different materials such as polyester, ceramics, aluminum, mica, etc. but the way they work remains the same.)

There are two main categories of capacitors to keep in mind during testing. There are polarized and non-polarized types. **Polarized** means that the capacitor has a polarity, in other words a positive side and a negative side. This means that these capacitors have to be installed in a circuit and/or charged a certain way depending on which side is positive and which side is negative. **Non-Polarized** means that the device has no polarity, therefore it can be installed either way or still work. Usually the way you will know if a capacitor is polarized is when one of the legs will be shorter than the other. The shorter leg is usually marked as the negative side and must be installed towards the negative of the circuit.

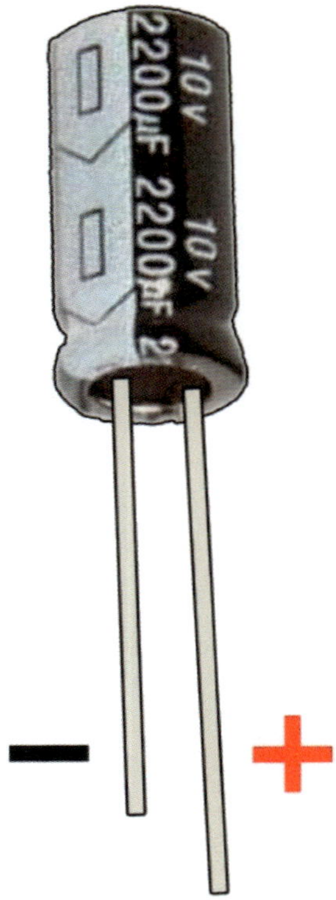

(Example of a polarized capacitor that has a positive side and a negative side. In this capacitor the way that you connect to it matters.)

Unit of Measure For Capacitance:

All capacitors are labeled in a unit called **Farads**. When testing for capacitance, you should expect to get a farad (*f*), milli farad (m*f*), micro farad (u*f*) and even a pico farad (p*f*) reading on the multimeter screen.

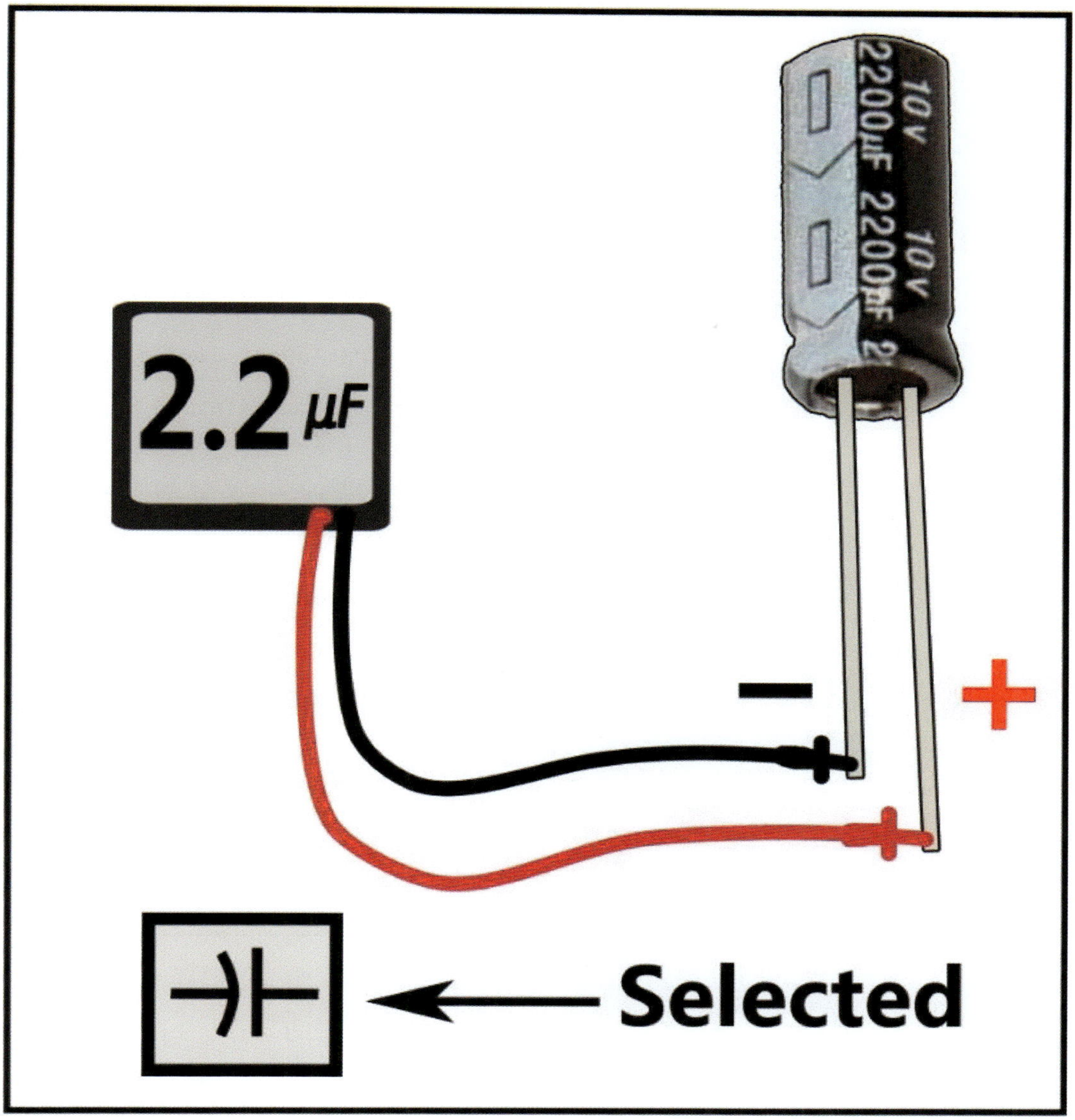

(To test a capacitor for capacitance, set the meter to the symbol for capacitance. Touch the two test probes to the legs of the capacitor. Make sure to connect positive with positive and negative with negative, if the capacitor is the polarized type,such as the one in this example. You should get a reading in the unit **farads** that matches what the capacitor has stamped on its case.)

Note: The capacitance test on the multimeter is usually only accurate to small capacitors. It loses its accuracy on bigger capacitors. To test any capacitor more accurately, an actual **capacitance meter** that is designed only to test capacitors should be used on bigger sized capacitors.

Ch.5: Hz & Duty Cycle

These meter settings are unique to only being used during special situations. These setting may or may not ever be used and are usually used to gather further information about how an electric circuit is working. It is usually more commonly used during the designing process of a circuit. You may only use these setting when building electronics or when testing for quick electrical signals from sensors but even then the preferred method of testing is to use a different test equipment known as an oscilloscope or a graphing multimeter. Nonetheless, I will explain how this function works on the basic multimeter.

What Is [Frequency/Hz]:

Frequency can be best explained as a measurement of how frequent per second electricity changes in voltage level. Up until this point we have only seen a circuit be either ON or OFF. However, in many of today's circuits involving electronics, the electricity is switched ON and OFF very fast. The idea behind this is that by switching something electrical on and off very fast you are able to still power the device the same as if it were ALWAYS ON, but also save electricity from when it is in the OFF part of the switching. This varying electricity is commonly achieved by a different kind of power source called a switching power supply.

Let's take a look at a graph showing two ways how electricity levels might change over time..

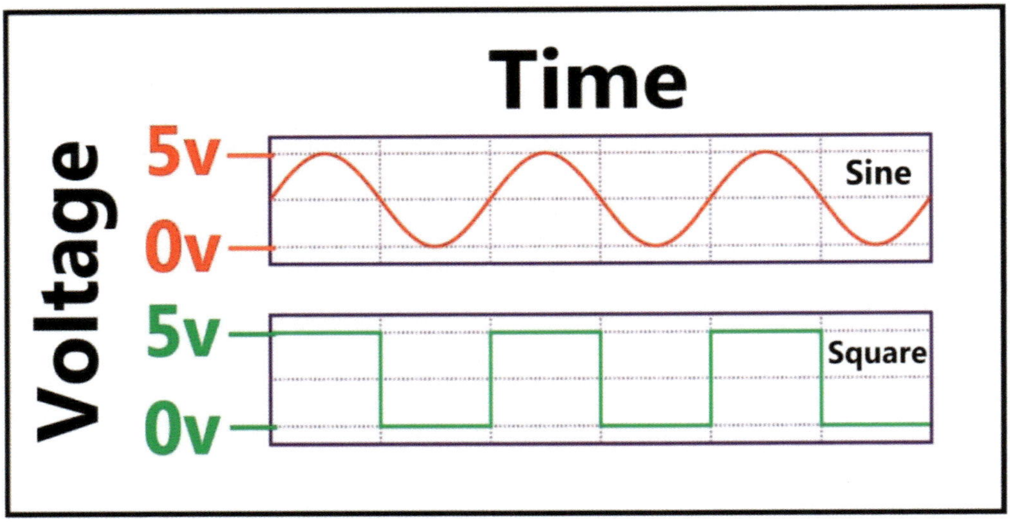

(Different styles of switching electricity and what they would look like when graphed.This is only for educational purposes to explain how the voltage level changes during switching. None of these graphs will show up on a normal multimeter when testing. The red graph represents an AC sinewave pattern while the pattern in green is of a DC square wave.)

The switching of electricity is done very fast and can only be noticed using a meter to measure it. The unit for measuring the rate that electricity is switching (frequency in a circuit) is called **Hertz(Hz)**. The most common frequency I can think of is the 60hz from your household power outlets. The 60hz means the electricity in the circuit is being switched on and off 60 times per second!!

Hertz: The unit displayed when measuring frequency in a switching circuit. This is a measure of how many times per second the voltage level changes from one level to another.

Example 1: If a circuit has a frequency of 2000 hertz, this means that the voltage level changes from high to low 2,000 times per second!

Example 2: If a circuit has a frequency of 200k hertz, this means that the voltage level changes from high to low 200,000 times per second! Pay attention to the K before Hertz in this second example. The "K" after a meter reading means thousand.

How To Measure Frequency:

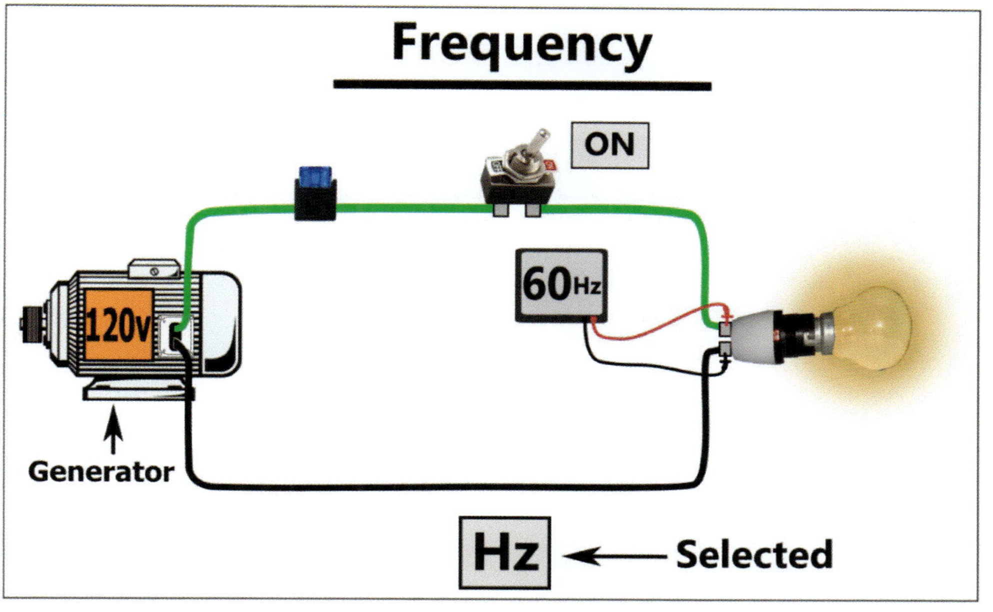

(The method for measuring frequency is just like you would do using the voltmeter. You must touch the tip of the test probes to the inside of the wire or use the terminals of the electrical device as test points. You can either test at the electrical load (as seen here) or at the power source. With the multi-meter set to (Hz) you will be able to read how often per second the electricity switches. The frequency of the circuit will be displayed in a number value on the meter screen. The reading on the meter shown is 60hz meaning it switches from high voltage to low voltage 60 times per second.)

Note: The Frequency setting can only be used on a circuit that switches the power on and off very fast. A normal everyday circuit that doesn't switch will read nothing if you try to use these settings on it.

Let's take a look at examples of what frequencies can look like visually.

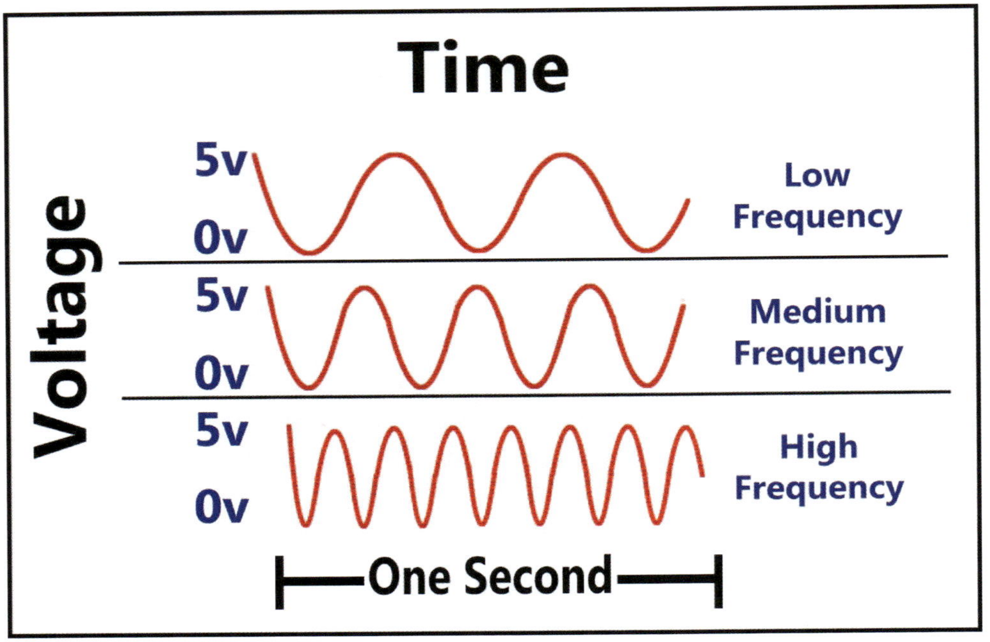

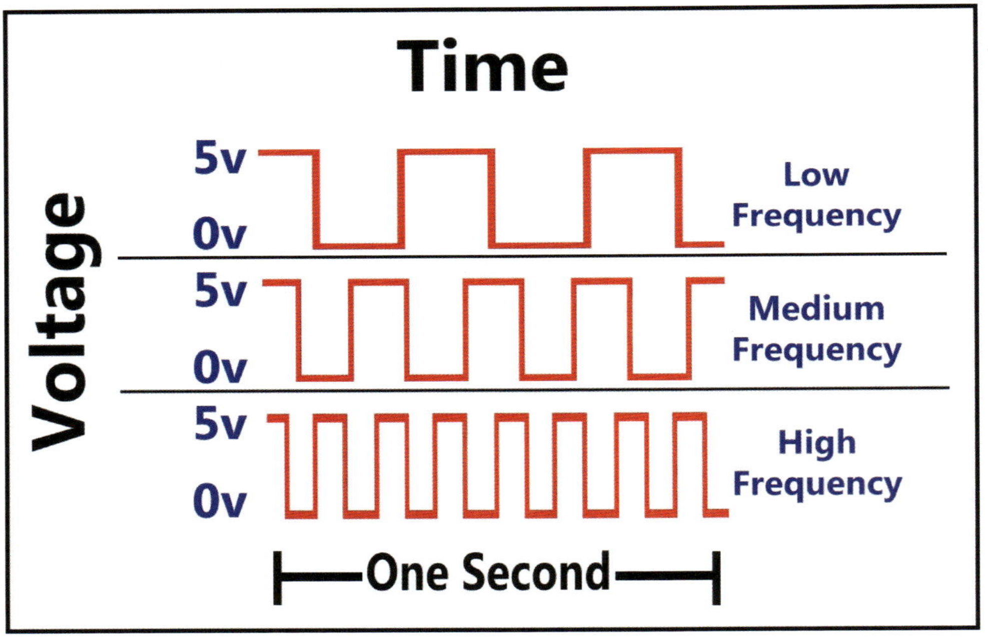

(This example is a graphical view of what is actually happening to the voltage levels when referring to frequency. As you see here, the voltage in this example is switching from 5 volts to 0 volts. The amount of times it switches high to low per one second is the frequency. The more switching per second, the higher the frequency.)

Cycles:

A cycle can be explained as one completion of going from a high voltage to low voltage, once. Each cycle is equal to one hertz. The amount of cycles completed per second is the **frequency** of the electricity.

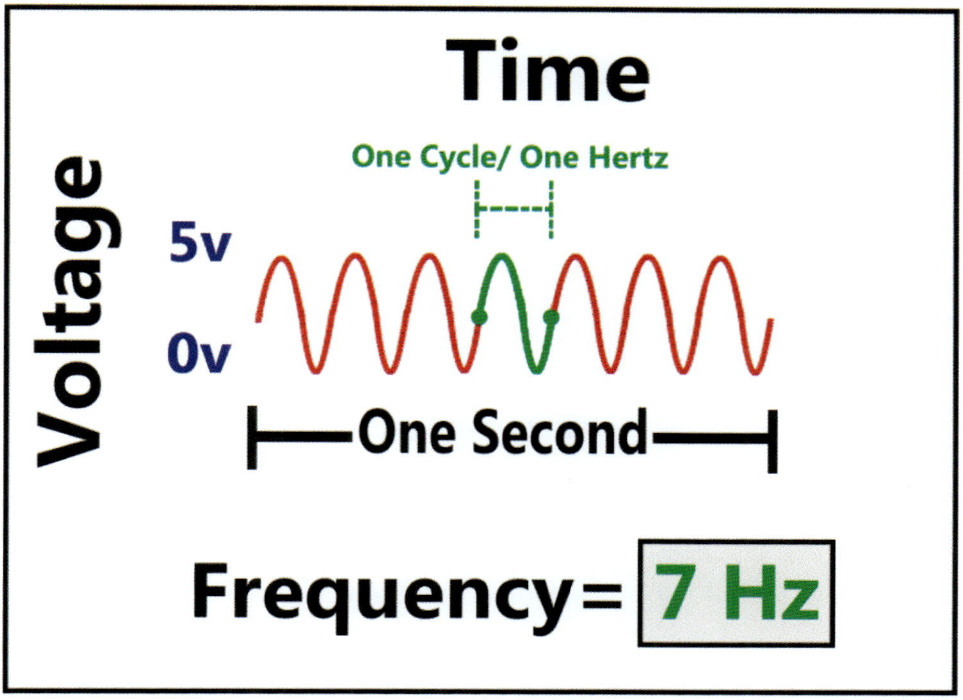

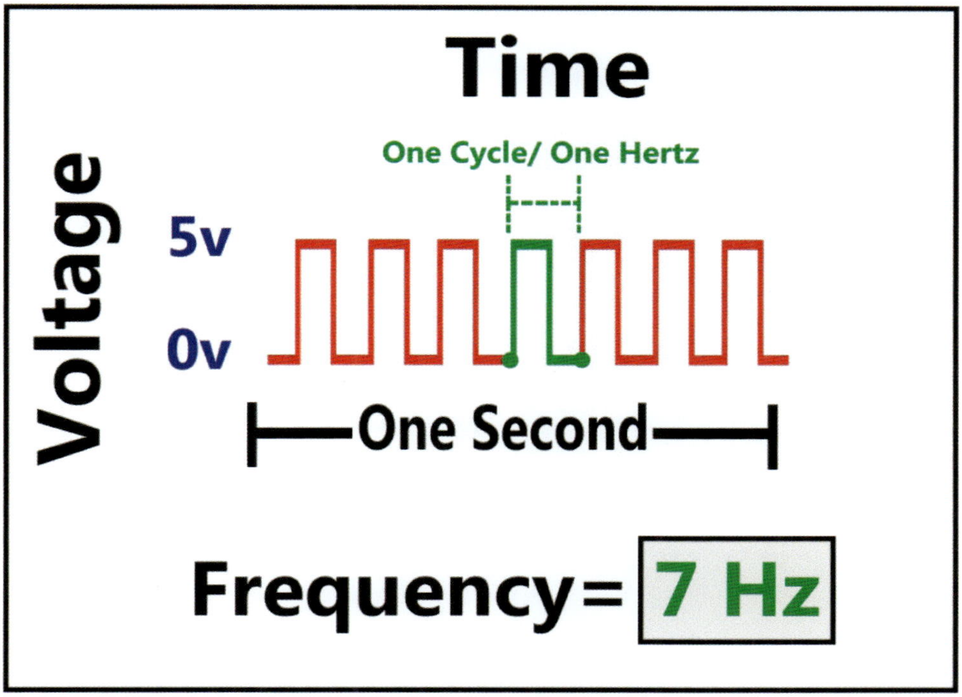

(More examples of cycles and frequency. A cycle is one complete transition from high to low. The amount of times per second the electricity switches from high to low voltage per second is the frequency. In the first example the frequency is 7 Hz. There are 7 sine wave cycles completed in one second so the frequency is 7 Hz. In the second example the frequency is also 7 Hz. There are 7 square wave cycles completed in one second so the frequency is 7 Hz.)

Note that the normal multimeter **WILL NOT** display this switching of voltage on the screen, as I have been illustrating here. Your meter will only display the number value of the frequency. I have drawn these graphs to explain what you are actually measuring.

Duty Cycle:

The duty cycle of a circuit is related to the talk about frequency. Duty cycle is a **percentage reading** calculated from the amount of "on time" as compared to "off time" of the switching voltage levels.

Note: To calculate this measurement manually you can take one cycle from the frequency and compare the High (ON) portion to the Low (OFF) portion. This can only be visually performed using an **Oscilloscope** or **Graphing Multimeter**. I will give more about how to use this other test equipment in another book. For now, let us continue with the explanation on duty cycle.

Frequency and duty cycle are very often confused to be the same. They are not the same. Remember frequency is how frequent per second the electricity is changing. Duty cycle is a **percentage** of the on time compared to the off time of one cycle.

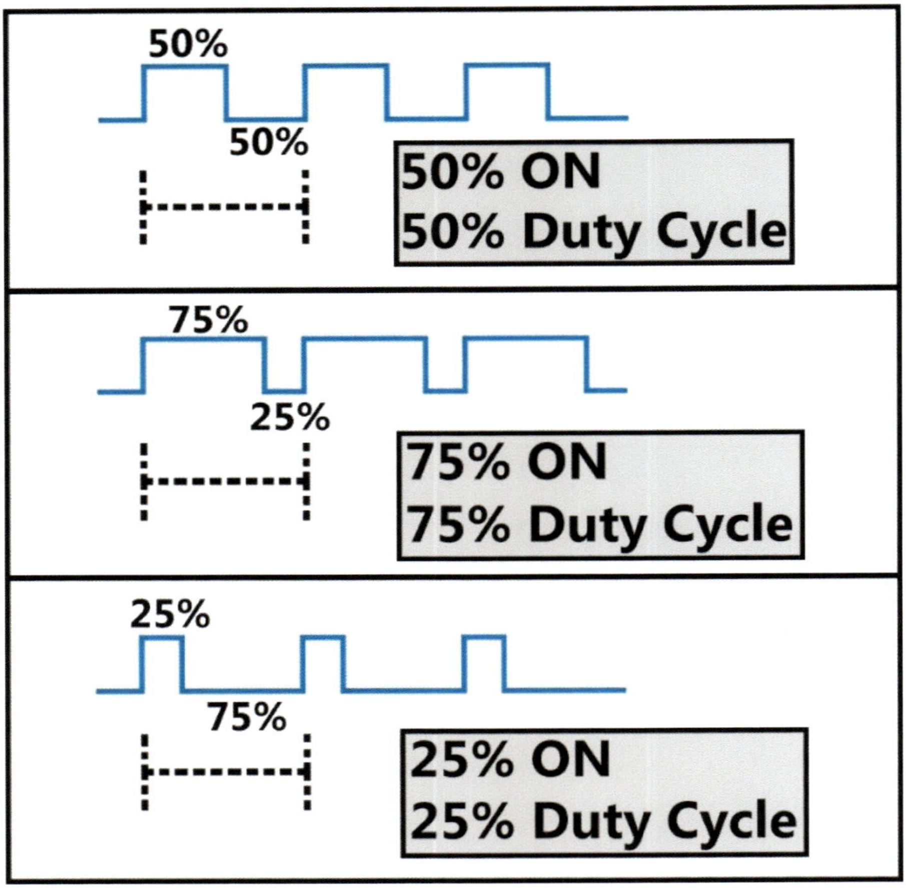

(Example of Duty Cycle: The percentage of on time, compared to off time of one cycle. The longer the ON time compared to the OFF time, the higher the duty cycle percentage displayed on your multi-meter. In this illustration there are three examples of duty cycles and what they actually look like when graphed. The percentage for duty cycle is calculated from the ON time when compared to the OFF time of one cycle.)

Measuring Duty Cycle:

Now I must admit these dial settings may or may not ever become useful to you depending on what you are testing but regardless I must show how to test for it simply because it is a very common feature for a multi-meter to have.

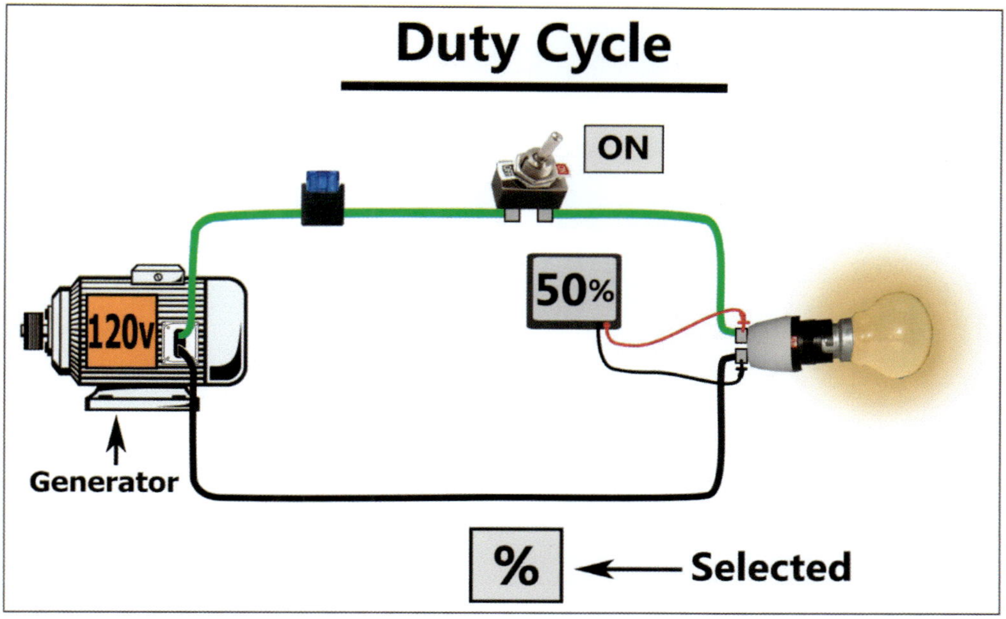

(Testing for duty cycle is done just as you would do with a voltmeter. You must touch the tip of the test probes to the inside of the wire or use terminals of the electrical device as test points. With the multi-meter set to (Duty) you will be able to read the percentage of on time the electricity switches. The reading on the meter shown is 50% meaning the ON time of the voltage is 50% compared to the off time.)

Note: The Duty Cycle setting can only be used on a circuit that switches the power on and off very fast. A normal everyday circuit that doesn't switch will read nothing if you try to use these settings on it.

Ch.6: Temperature

The final dial setting you may have on your meter is the Temp setting. This is setting is very self-explanatory. You can measure the temperature of a surface using the special test probe provided. In this test setting you remove your normal test probes and install a temperature probe in their place.

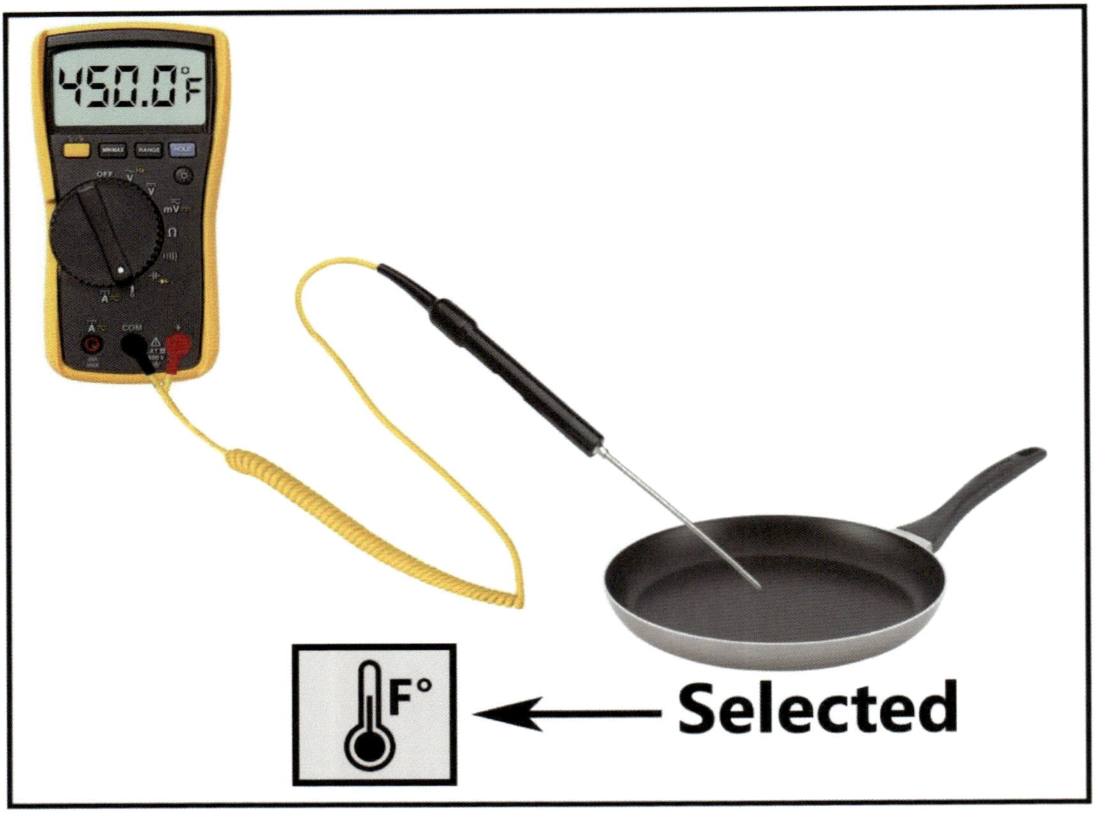

(This is an example of a temperature probe usually provided with the purchase of a multi-meter with a TEMP setting. The probe goes in the place of the normal red and black test probes. Once the dial setting is on temperature and the probe is connected, it is simply a matter of touching something with the end of the probe to get a temperature reading off it.)

This setting is probably the easiest to use. The only thing I have noticed about the temperature probe is the fact that many will not measure a temperature unless it is physical touching an object. That means you might not be able to measure the outside air temperature with it if you wish to do so.

Ch.7 Graphing Multimeters and Uses:

The graphing multimeter is a special kind of multimeter that has one additional function of being able to graphing the test results instead of just reading a number value. It works just like a normal multimeter would only now the screen can display the readings in either digital number form or in a graph that shows changes in measurements over time.

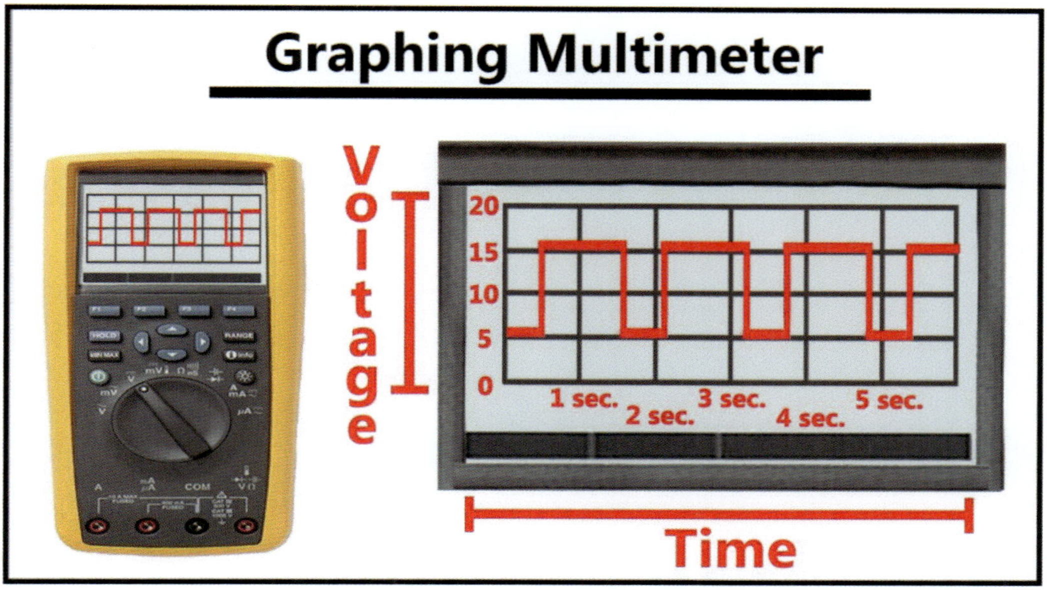

(This image shows an example of a graphing multimeter in **Graph** setting, measuring a changing voltage. The vertical scale is the Voltage scale and the horizontal scale is the Time scale. The display on this graphing meter shows a square wave signal that is switching in voltage from about 16v to 0v over a period of 5 seconds.)

The graphing multimeter is essentially hooked up to a circuit the same way as the normal multimeter would only now you will have the option of graphing your voltages and amperages.

AC Electricity Graph View:

Let's take a look at what the graphing multimeter would display for the voltage coming from a generator..

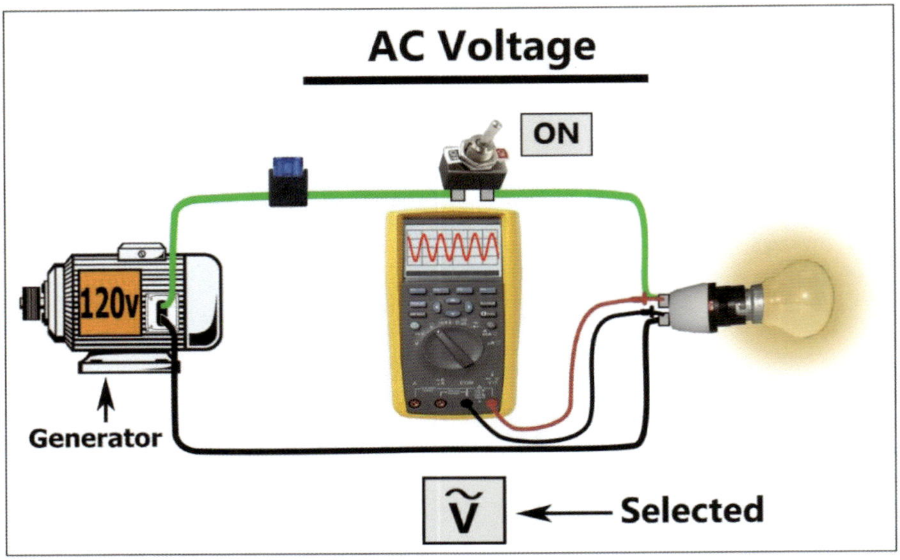

(Example of the graphing multimeter in graph mode showing the Alternating Voltage coming out of an AC generator. The voltage reading looks like a series of waves, gradually increasing and decreasing. This is known as a **sinewave.**)

In the previous example we notice the varying sinewave that is displayed by the meter when measuring an AC voltage. This is represented on the screen by a voltage reading that is gradually going up and down like a series of waves.

This additional GRAPH function can be useful because of the fact that it can graph voltages or amperages that change over time. One of the previous chapters dealt with the Frequency and Duty Cycle function using the basic multimeter. With the graphing multimeter, it is now possible to visually see both of these things in action.

DC Electricity Graphed:

Now let's see how a constant DC voltage would be displayed when using a graph meter.

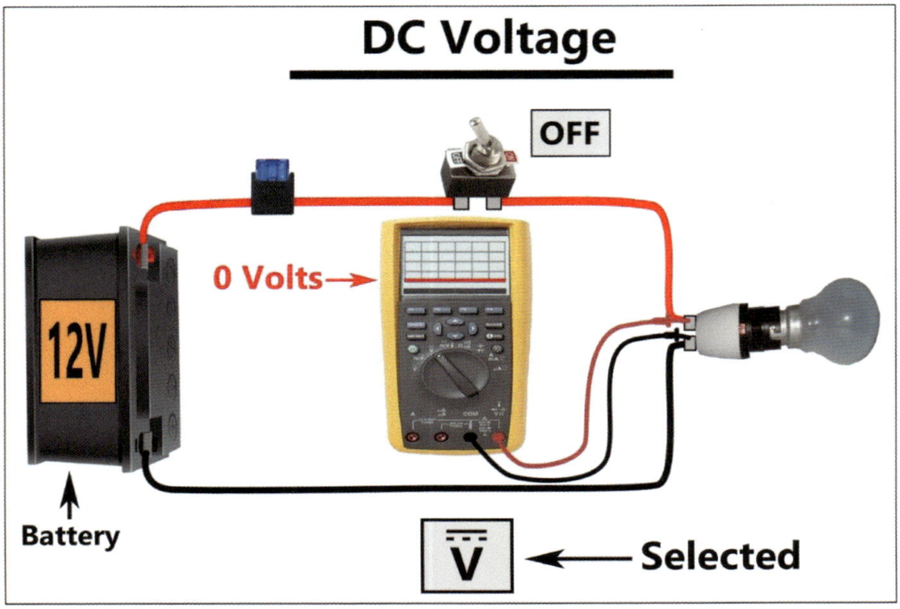

(This image show the graphed readings of this meter with the circuit turned OFF. The reading is displayed as a solid horizontal line in the vertical scale that matches the voltage being measured. This reading indicates 0 volts.)

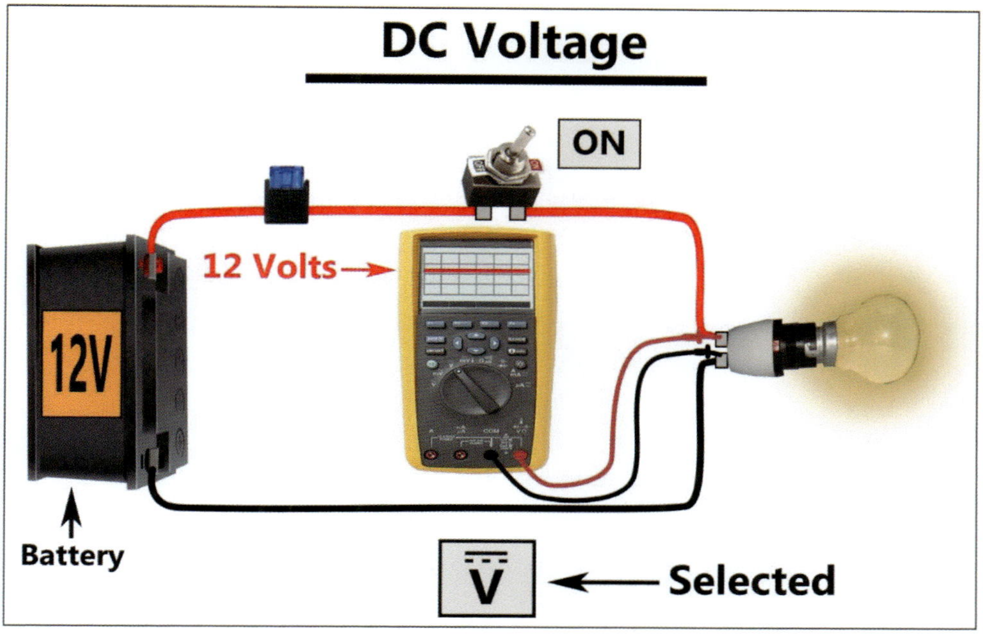

(This image shows the graphed readings with the circuit turned ON. The reading is displayed as a solid horizontal line in the vertical scale that matches the voltage being measured. This reading indicates 12 volts.)

In these illustrations we see an example of the graphing meter being used to measure a DC voltage. We have sampled a voltage reading for when the switch is OFF and also when the switch is ON. The meter will display a solid horizontal line in the vertical level that matches the voltage being measured.

This same idea can be applied to AC or DC amperage readings except that hook up to the test area would be done how you would normally hook up an ampmeter. If you were measuring AC amps, the graphing meter would display a varying amperage just like in the AC voltage example. If you were measuring DC amps, the graphing meter would display a horizontal line indicating constant amperage just like in the DC voltage example.

The Graphing Meter for Intermittent Faults:

One other great use for the graphing multimeter is the fact that the graph can help tremendously when trying to find intermittent electrical problems. By intermittent I mean that the problem does not always happen. The voltage to the circuit might cut off only temporarily and causes a problem. Then all of a sudden the circuit may start to work again. This can happen because of a loose connection somewhere in the circuit that temporarily makes and breaks electrical contact. These are some of the hardest electrical problems to track down but with the graphing multimeter, it would be a lot easier to notice. The benefit to using a graphing multimeter is the fact that it is faster and also able to catch certain drop outs of voltage that last only milliseconds that would not be seen using a regular multimeter. Let's take a look at what an intermittent problem looks like in action..

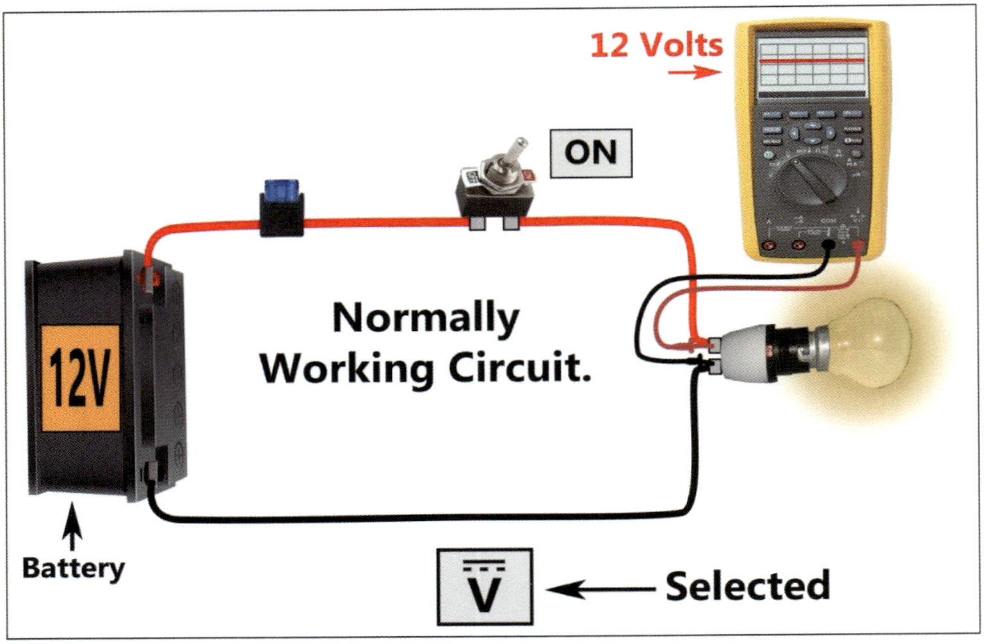

(Normally operating circuit. Light is ON and everything is working properly. The meter reads a 12 volt flatline)

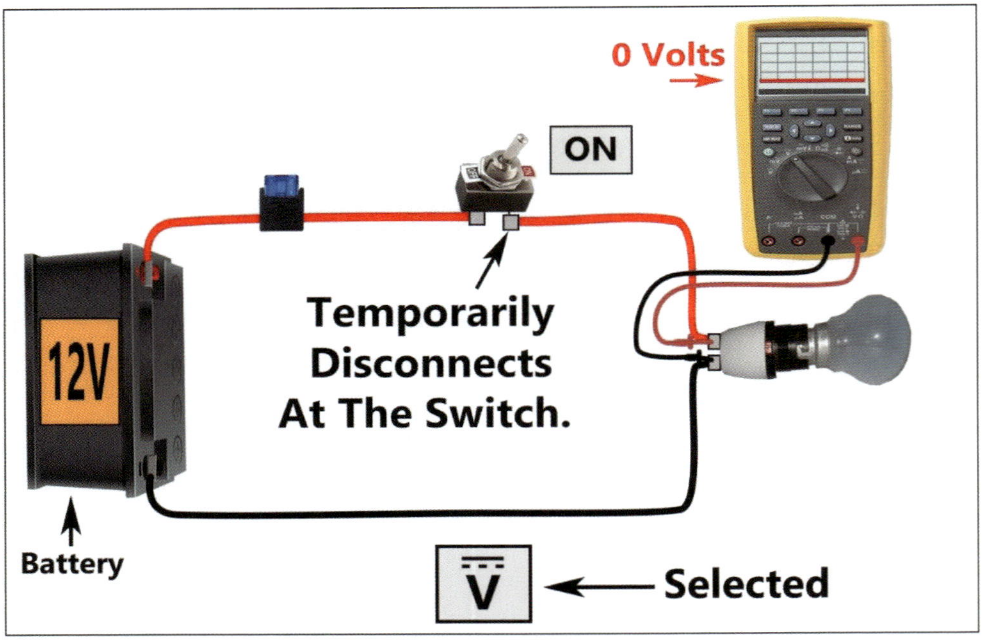

(The circuit cuts OFF intermittently due to a loose connection at the switch. The meter now reads a 0 volt flat line.)

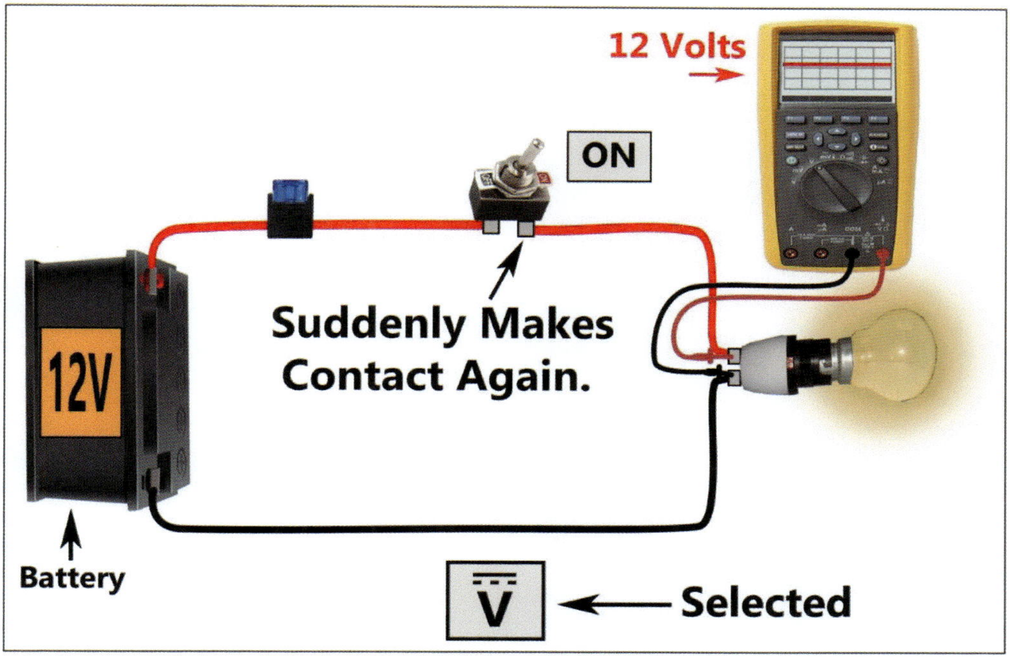

(While further testing and wiggling of the wires to check for tightness the connection at the switch made contact again and the light turned back ON. The meter reading 12 volts again.)

In these illustrations we see an example of an intermittent electrical issue being monitored using the graphing meter. By probing at the light bulb we can notice how the voltage drops out to 0 volts when the intermittent problem is happening. This happens when there is a loose connection somewhere in the circuit that can break and remake contact easily.

Further diagnosis would be necessary for this kind of problem. I have included a variety of test methods for these kind of harder problems and many more in my other book "How To Test Like A Pro: Part 2"

Note on Graphing Multimeter Limitations:

The graphing multimeter does have its limits. The only problem with the graphing meter is that it can generally only be used for graphing low speed changes in voltage or current. By low speed I mean that even the best graphing multimeters are limited to only be able to catch switching in electricity that last about 1/10 of a second (100milliseconds). Many others can only display readings that last 1/2 a second and the average graphing meter is only able to catch voltage changes that last a second.

This meter is **NOT** suitable for measuring high speed voltage signals that last only one thousandth of a second and even faster signals that last one millionth of a second. Such high speed signals would be like the ones from switching power supplies, computer signals and electrical sensors. To be able to measure these kinds of fast signal you will need to use faster test equipment.

Ch.8: Multimeter Accuracy and Choosing the Right Meter:

In order to properly test any electrical circuit or electrical device you must first be 100% sure that the meter that you are using is accurate enough to give correct values. If the meter you are using for testing is inaccurate enough, the readings on the meter might be falsely misinterpreted as an electrical problem with the circuit. This can be very confusing, especially for a beginner in electricity because the reading on the meter might be too far off from the real reading. When buying a multimeter, the accuracy percentages for each dial setting on the meter are given in the spec sheet for the product. It is important that you know exactly how off your multimeter is from the true readings in order to not misinterpret any measurements that you perform.

Measurement Unit	Range	Accurancy
DCV	0~1000V	±(0.03% +5)
ACV	0~700V	±(1% +5)
DC mA	0~40mA	±(0.05% +2)
AC mA	0~40mA	±(0.05% +2)
DC A	0~10A	±(0.8% +40)
AC A	0~10A	±(0.8% +40)
RESISTANCE	0~40MΩ	±(0.1% +6)
FREQUENCY	0~50KHz	±(0.05% +5)
DIODE TEST	0~2.4V (Shows Diode Voltage Drop) 2% +1	
CONTINUITY	Beeps for Resistance < approx. 90 ohms	
BATTERY TYPE	x6 1.2v Ni-MH Battery	

(In this example we see a spec sheet chart for a multimeter, which includes all of its accuracy ratings. The percentages for how much this meter's readings may vary from the real reading is listed by measurement function. Each dial function of the multimeter has its own accuracy percentage.)

From each accuracy percentage you can calculate the maximum amount that a reading might vary, higher or lower compared to the real reading. Let's take a look at how to calculate that now...

DCV = ±(0.03% +5)

Step 1: Convert Percentage To Decimal.

Example For Percentage To Decimal:
1%= 0.01 in decimal form.
So
0.03%= 0.0003 in decimal form.

Step 2: Calculate The Accuracy.

The True Voltage Reading is 12v.
So
(12v) x (0.0003)= ±0.0036v
So
The meter reading will vary
From 12v + 0.0036v= 12.0036v
To 12v - 0.0036v= 11.9964v

Step 3: Add/Subtract (5) To Last Digit.

12.0036v +(5 to last digit)= 12.0041v
11.9964v +(5 to last digit)= 11.9959v

Final Results: From 12.0041v to 11.9959v.

(In this illustration we see the steps to calculating the variation in readings of a meter that has a **(0.03% + 5)** DC voltage accuracy rating. The first step is to convert the percentage rating into a decimal. Then we multiply this decimal with the true voltage to get our ,plus or minus, voltage variation values. Next we add/subtract this voltage variation value from our true voltage and we will end up with two numbers. A high end variation and a low end variation. Last, we add/subtract the second number in our original accuracy rating to these two numbers we got from calculations. The final two result numbers is the range at which the meter readings will vary from the true voltage reading. In this example the range of variation is **11.9959v to 12.0036v**. This is not too far off from the true 12v reading so this meter would be considered to have good accuracy for DC voltage.)

Now lets take a look at another example of an accuracy rating.

DCV = ±(3% +5)

Step 1: Convert Percentage To Decimal.

Example For Percentage To Decimal:
1%= 0.01 in decimal form.
So
3%= 0.03 in decimal form.

Step 2: Calculate The Accuracy.

The True Voltage Reading is 12v.
So
(12v) x (0.03)= ±0.36v
So
The meter reading will vary
From 12v + 0.36v= 12.36v
To 12v - 0.36v= 11.64v

Step 3: Add/Subtract (5) To Last Digit.

12.36v +(5 to last digit)= 12.41v
11.64v +(5 to last digit)= 11.59v

Final Results: From 12.41v to 11.59v.

(In this illustration we see the steps to calculating the variation in readings of a meter that has a **(3% + 5)** DC voltage accuracy rating. Again, the first step is to convert the percentage rating into a decimal. Then we multiply this decimal with the true voltage to get our ,plus or minus, voltage variation values. Next we add/subtract this voltage variation value from our true voltage and we will end up with two numbers. A high end variation and a low end variation. Last, we add/subtract the second number in our original accuracy rating to these two numbers we got from calculations. The final two resulting numbers is the range at which the meter readings will vary from the true voltage reading. In this example the range of variation is **11.59v to 12.36v**. This is way off from the true 12v reading so this meter would be considered to have bad accuracy for DC voltage.)

The second meter's accuracy rating was way too inaccurate for DC volts. When you are doing precision electrical testing, a meter with this kind of accuracy would lead you to believe that there was problem somewhere in the circuit if you read 11.64v. This is the reason why knowing the accuracy of your multimeter matters.

The goal when purchasing a multimeter is to obviously get the best deal for the money. You can accomplish this goal by making sure that it has all the most important testing functions on the dial of a multimeter. At the same time, you also MUST check the accuracy rating for each of the dial functions to determine if the meter you are looking to buy is right for the application you will be using it for. Let's take a look at some examples of how three different meters with different accuracies may vary from the real measurement reading...

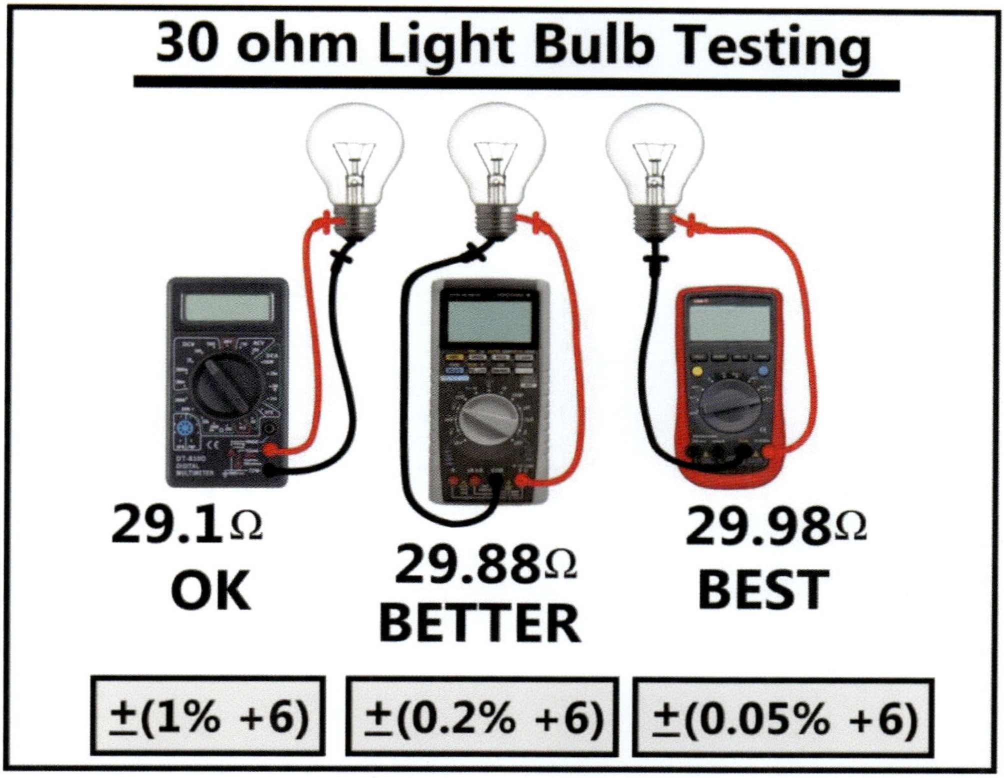

(In this illustration there are three different multimeters with different accuracy ratings. They are all put to the test to measure resistance to see how they perform testing the same electrical device. The three displayed three different resistances. The best reading was from the meter with the lowest percentage rating. This meter was the closest to the actual true reading.)

The smaller the percentage number on the accuracy rating, the better the accuracy is. When choosing a multimeter always take into account the accuracy so you will never have to question whether a problem really exists or not. With this information, you should be able to choose the best multimeter for your specific application.

Ch. 9: Miscellaneous Tips And Lessons

In this chapter I will include some of the common beginner mistake to look out for as well as new tips and tricks to make electrical testing easier. These lessons are an add-on to take with you on your journey to electrical testing. Enjoy.

Using Contacts as Test Points:

In some cases, there may be circuits that have contacts and terminals that are exposed and easily to accessible. If this is the case with the circuit you are testing one great thing about this is that you can actually use contacts as test points for voltage measurements if they are easy to access. This allows you to test at certain points of the circuit without having to pierce any wires.

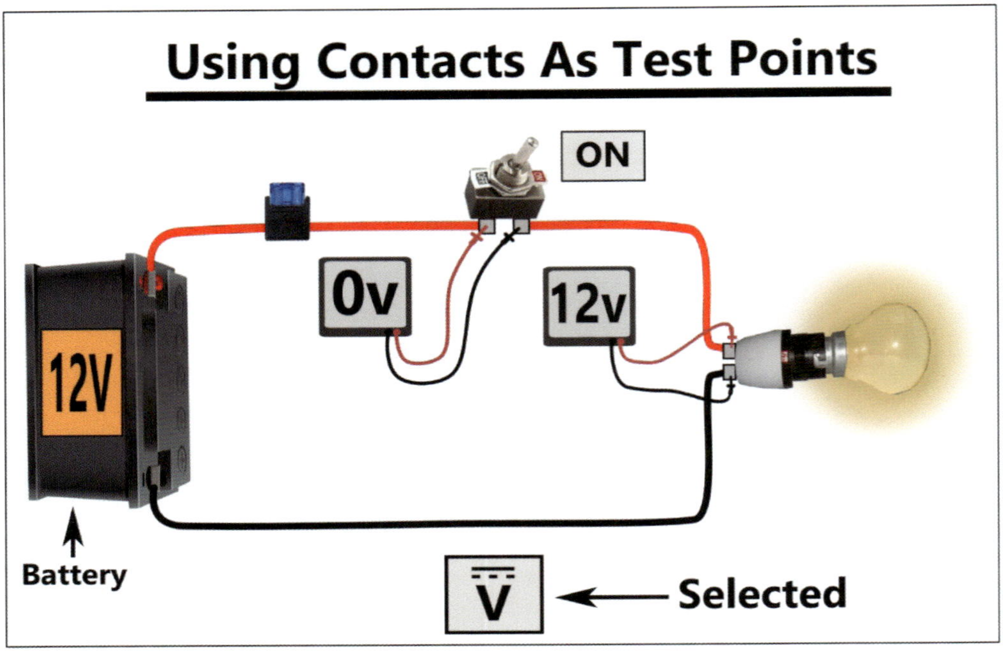

(This is an example of what a voltage test looks like using the contact points where the wiring would connect to parts of the circuit like a switch or a light bulb. Remember that you must have the circuit ON for proper voltage testing.)

Using Push Pins or Sewing Needles For Backprobing:

For the electrician who prefers not to poke wires then this may be the method for you. Testing at connectors can be easily done by slipping in a long push pin or thin needle through the back of the connector to create a test point. The trick to this method is to make sure the tip of the pin is contacting the metal terminal on the inside of the connector. If you don't contact the metal terminal inside the connector you will not read anything when it's time to use it as a test point for voltage measurements. This method takes a little getting used to in order to tell if the t-pin is contacting the inside or not.

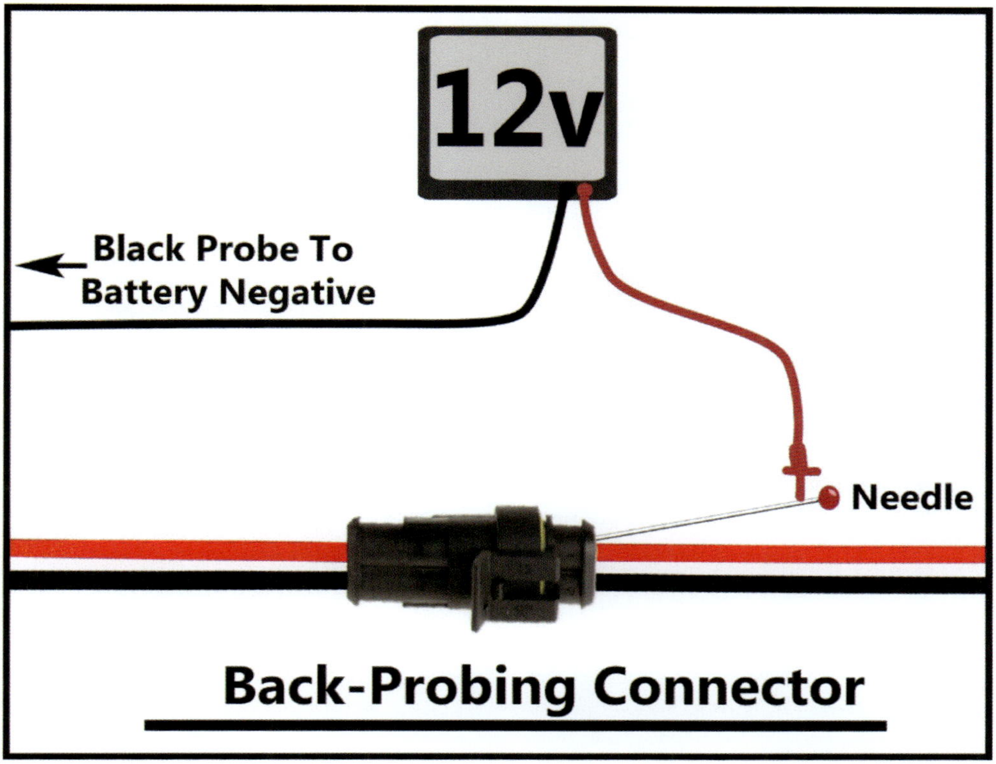

(This image shows a needle being used to back probe the connector in order to get a voltage reading while its still plugged in. The trick to this method is to make sure the needle is contacting the terminal inside of the connector.)

Test Probes and Accidentally Creating a Short During Testing:

One thing to understand and remember is that when you have a power wire and ground wire very close to each other there is always a potential for a short to happen. The most common area where you might accidentally cause a short is at the electrical load of the circuit (In this case at the light bulb). In a best case scenario if you create a short you will blow the circuit's fuse and the circuit will not work anymore until the fuse is replaced. The worst case scenario is if the circuit is not fused then you may you accidentally damage a part of the circuit.

Back Probing Mistakes:

Many times during testing, the contacts and terminals of the parts in the circuit are not exposed and therefore are difficult to use as test points. instead of using a wire piercing probe during testing, technicians may prefer using regular t-pins as a cheap replacement for insulated back probes to test a circuit. This practice is okay to do if it is easier, as long as the t-pins do not touch and create a short during testing. If one T-pin is probed to power and another is probed to ground keep them away from each other as you may accidentally blow a fuse by letting them touch.

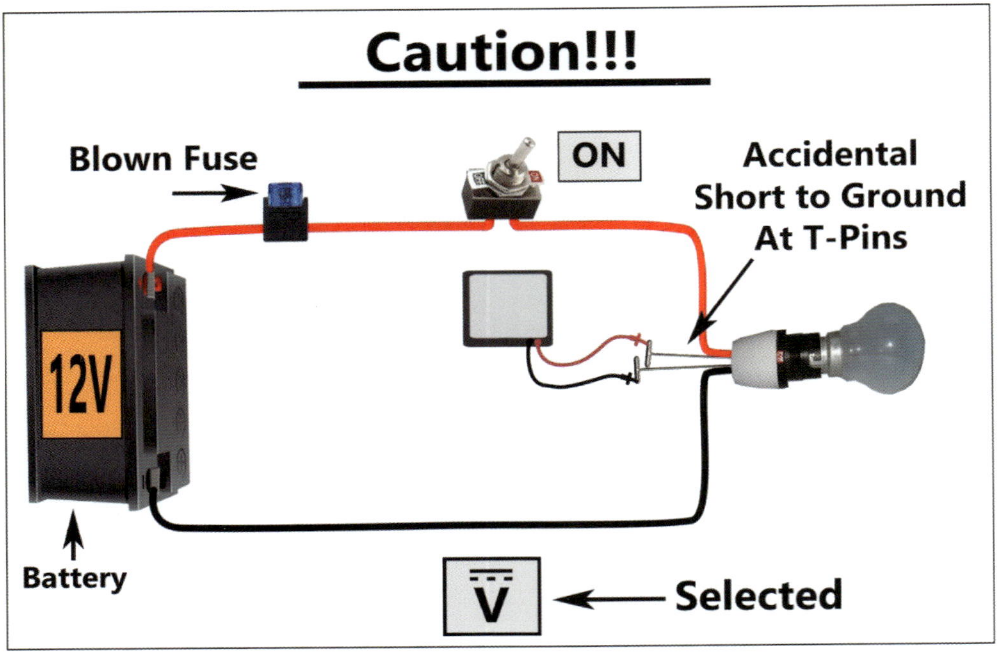

(Example of a circuit that doesn't have contacts readily available to use as test points. The technician preferred to use t-pins to probe the backside of the light bulb. The t-pins accidentally touched and blew the fuse of the circuit which in turn caused the light bulb to go out. In this case the repair for the mistake was to simply replace the blown fuse.)

Remember that if one T-pin is probed to power and another is probed to negative keep them away from each other to prevent an unintentional electric short. One way to prevent this is to use insulated back probes or to insulate the t-pins being used as probes.

Accidentally Crossing The Test Probe Tips During Probing:

Another way to accidentally created a short is by crossing the test probes themselves during testing. Sometimes to check for voltage at the electrical load, it is a lot easier to just disconnect the device and then test at the terminals for voltage. Depending on the type of connector or terminals, there may be a very limited space for the test probes to fit in. Care must be taken when trying to probe the front of a connector.

A problem I caused once was when I stuck the test probes to each of the light's terminals. There was very limited space and I was trying to get a quick reading to confirm voltage. What I didn't notice was that the two test probes were touching each other at the same time they were touching the pins in the connector. One pin had 12v and the other pin had 0v so by accidentally bridging the two pins using my probe tips I effectively created a short to ground in the circuit.

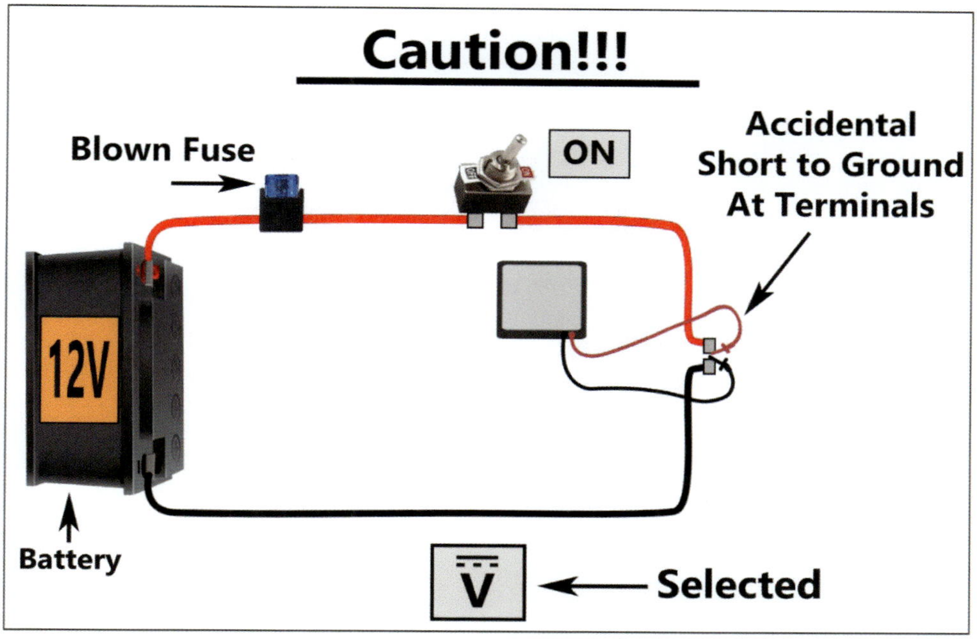

(Example of an accidental short created by crossing the two test probes while inside of a connector with limited space. The connector had positive in one pin and negative on the other pin. By crossing the test probes and touching the pins at the same time a short to ground was created. Care must be take if you are front probing a connector with limited space between its pins.

(It is especially easy to make this mistake and create a short with the terminals inside of a connectors like the ones shown here. There is no insulated separation between the pins inside of these types of connectors.)

These mistakes of accidentally creating shorts have far worst consequences when testing circuits with computers and control modules as you can potentially damage one of these computers. This can cost you thousands in expensive replacements parts so be careful. The multimeter's test probes are made of metal therefore they are conductive and can create a short if you are not careful during testing. Do not let the probes touch each other when they are already touching a set of live terminals.

Conclusion: (Summary & Ending Words)

Thank you very much for purchasing my book and I really hope that you have learned something valuable from me. I work hard on these illustrations and explanation and I hope that I have taught you something you can actually use. This is only the beginning and I hope to see you soon in the next books of my series "Everything Electrical."

In this book I only aimed to teach the basics of how to use each meter setting. There are far more skills you will come to learn that will make you very good at the settings of the multi-meter and using them to troubleshoot the various electrical issues that exist. I have made more books to cover all the advanced subject and issues more in depth. Keep an eye out for new books explaining other electrical subjects you might feel confused about. Stay Tuned and Thank you for reading.Good luck.

Printed in Great Britain
by Amazon

43050810R00048